PRAISE

ADVENT OF THI

"Here's a foolproof mathematica
EB. Translation: David Berlinski
traordinary book. [He] is one of those rare authors who know the information culture from the inside out. . . . Making simple and accessible that which had previously been murky and intimidating is Berlinski's specialty." —*Chicago Tribune*

"In short, densely intertwined, lyrically constructed chapters, Berlinski describes the discoveries of major algorithmic thinkers. . . . A novelist as well as a mathematician, Berlinski has composed energetic, intertwined tales that make it nearly impossible for readers, once drawn in, to lose interest or to get lost among flying abstractions. . . . An uncommon achievement of both style and substance."

—*Publishers Weekly* (starred review)

"For Berlinski, the tortured history of the algorithm illuminates more than just the tethered reach of logic: the thinkers responsible for its development stride across his pages as real people, their frustrations and hopes spilling over into the fictional interludes interspersed throughout the book. A tour de force, this book gives intellectual dilemmas a human face, while restoring grandeur and mystery to a universe still too richly intricate to fit within a computer protocol."

—*Booklist* (starred review)

"Engaging . . . Interspersed throughout the book are little asides of easy reading. They may be sensitively written vignettes from the lives of the mathematicians he is writing about, autobiographical episodes from his own life, fables or fanciful stories. These interludes . . . help to give some insight as to why so many brilliant people have devoted their lives to this strange field." —*The Washington Post Book World*

"[A] deep and instructive account of the algorithm's development and its role in modern life. . . . Rewarding information about mathematics, famous and not so famous mathematicians, and philosophy." —*Scientific American*

"In his most recent book, David Berlinski is by turns a mathematician, a storyteller, a would-be poet, and a philosopher. . . . Touching upon disciplines as varied as physics, mathematical logic, and psychology, [he] carries the reader from anecdote to technical explanation to philosophical speculation to lyrical musing and back again with barely a pause. . . . *The Advent of the Algorithm* is not an academic book by any stretch, but neither is it a typical example of popular science-writing, being designed as much to excite the imagination as to inform. It does both, and then some." —*Commentary*

"[Berlinski] succeeds in carrying his readers through the basic notation of mathematical logic in a fashion that should work well even for lay readers. [A] zesty and unusual book."
—*Library Journal*

"Berlinski imparts an understanding of the functional qualities of algorithms and how logicians such as Alan Turing, Kurt Gödel, and Emil Post initiated the widespread use of algorithms, which now extends beyond computers and into the realms of biology and social organization." —*Science News*

"Berlinski brandishes his opinions fiercely and fearlessly. . . . He offers the impressive insights of original thinkers from a staggeringly wide variety of intellectual pursuits and persuasions, including, among a great many others, Immanuel Kant, René Descartes, John Dos Passos, Richard Feynman and Roger Penrose. —*Mercury News* (San Jose, CA)

"A scientific and cultural tour of the 350-year-old 'recipe' that made possible the current Digital Age." —*American Way*

The Advent of the Algorithm

Also by David Berlinski

On Systems Analysis: An Essay on the Limitations of Some Mathematical Methods in the Social, Political, and Biological Sciences

Black Mischief: Language, Life, Logic, & Luck

A Clean Sweep

Less than Meets the Eye

A Tour of the Calculus

The Body Shop

The Advent of the Algorithm

THE 300-YEAR JOURNEY FROM AN IDEA TO THE COMPUTER

David Berlinski

HARCOURT, INC.

San Diego

New York

London

www.harcourt.com

Library of Congress Cataloging-in-Publication Data
Berlinski, David.
The advent of the algorithm: the 300-year journey from an idea to the computer/David Berlinski.
p. cm.
Includes bibliographical references and index.
ISBN 0-15-100338-6
ISBN 0-15-601391-6 (pbk.)
1. Algorithms. I. Title.
QA9.58.B47 1999
511′.8—dc21 98-43755

Text set in Sabon
Printed in the United States of America
First Harvest edition 2001
J I H G F E D C B A

À la mémoire de mon ami

M. P. Schützenberger
1921–1996

A friend visited Dr. Johnson as he lay dying, and seeing that he lacked for support, placed a pillow beneath his head. "That will do," Dr. Johnson said, "all that a pillow can do."

AUTHOR'S NOTE

The Advent of the Algorithm contains a number of fictional episodes. These are set-off in the book by the following symbol: (■■■■). When I refer to these passages as "fictional," I mean that I invented them. I have met neither Isaac Bashevis Singer nor Jorge Luis Borges. Gottlob Frege and I did not team-teach logic together at one of the California community colleges. (The very thought is grotesque.) The Cardinal of Vienna never asked for my help in understanding the theory of recursive functions: I am entirely confident that he has no idea who I am. Throughout these sections, the first person voice is a narrative artifice. The exciting adventures undertaken by my namesake in this book were not undertaken by me.

Why these inventions?

And why not? Every writer, it seems to me, has the right to use the grand literary tradition as best he can. A great mathematical idea is a part of the community of human concerns, and some of the points I wished to make in discussing the advent of the algorithm could only be made obliquely, by means of reference to a world inhabited by suave Cardinals, mad Rabbis, Jewish dream merchants, and other characters who feel comfortable in such settings.

In this edition of *The Advent of the Algorithm,* I have corrected a number of misprints or blunders that appeared in the initial printing. In this respect, I am grateful to Linn Sennot and Ernest Nussbaum, who read what I originally wrote with enthusiasm and responded to its errors with alacrity. I am grateful as well to Professor Martin Davis for challenging my account of Alonzo Church's role in the creation of the calculus of lambda conversion. I had assigned credit to Church, and to Church alone, for proving that the lambda convertible and recursive functions are one and the same. The recognition, Davis argued, should go to Stephen Kleene. On reviewing the primary sources, I have concluded that, in fact, Church does deserve the credit I assigned him. But as Kleene established the same result very shortly afterward, I am more than happy to equally acknowledge these two great logicians. After all, there is more than enough credit to go around.

CONTENTS

PREFACE

The Digital Bureaucrat

More than sixty years ago, mathematical logicians, by defining precisely the concept of an algorithm, gave content to the ancient human idea of an effective calculation. Their definitions led to the creation of the digital computer, an interesting example of thought bending matter to its ends.

The first computer appeared in the 1940s and has like certain insects undergone pupation, materializing first as an oddity, and then in the 1950s and 1960s as a specter. In a famous cartoon from the *New Yorker,* a computer, when asked whether or not there was a God, responded that there was one *now.* Some sense remains that, like the sorcerer's apprentice, we have appropriated a device we do not understand and cannot control, but curiously enough, as the digital computer has become more powerful, it has also become less intimidating. After shedding several of its earlier incarnations, the computer has acquired the role it was destined all along to assume. It is essentially an enabling device, one serving to amplify the low babble of human needs, indispensable without being invaluable.

The digital computer is a machine and, like every material object, a captive in the end of various bleak laws of thermodynamics. Having run out of time, it runs out of steam. As does the computer programmer stabbing at the computer keyboard with the tips of two tense fingers. As do we all. An algorithm is otherwise. Occupying the space between the pinprick of desire and the resulting bubble of satisfaction, it is an abstract instrument of coordination, supplying procedural means to various ends. Contrived of signs and symbols, algorithms, like thoughts, reside in a world beyond time.

Every computer divides itself into its hardware and its software, the machine host to its algorithm, the human being to his mind. It is hardly surprising that men and women have done what computers now do long before computers could do anything at all. The dissociation between mind and matter in men and machines is very striking; it suggests that almost any stable and reliable organization of material objects can execute an algorithm and so come to command some form of intelligence.

In this regard, the apparatus of the modern digital computer is convenient but hardly necessary. To see this point is almost at once to see it confirmed. After all, what is a bureaucracy but a social organization that has since at least the time of the ancient Chinese patiently undertaken the execution of complicated algorithms?

If a bureaucracy resembles a computer at the level of social organization, the living cell, if it resembles anything in our experience, resembles a computer at the level of molecular organization. The metaphor is irresistible and few biologists have resisted it. And for good reason. No other metaphor conveys the intricacies of cellular replication, transcription, and translation; and, for all that we can tell, nothing besides an algorithm can handle the administration of the biological molecules.

These reflections might indicate that the digital computer represents less of a bright, bursting novelty in human experience than is generally imagined. Although true, this is, of course, a conclusion too reassuring to be completely true. There is a considerable difference between the execution of an algorithm by a social bureaucracy or even a bacterial cell and the execution of an algorithm by a digital computer. Having coaxed the concept of an algorithm into self-consciousness, the logicians have made possible the creation of algorithms of matchless power, elegance, concision, and reliability. A digital computer may well do what a bureaucracy has done, but it does it with astonishing speed, the digital computer possessing an altogether remarkable ability to compress the otherwise sluggish stream of time. This has made all the difference in the world.

Five hundred years have passed since Magellan sailed around the world, and the sun still comes up like thunder out of China in the east; yet the old heavy grunting physical sense of a world that must be physically circumnavigated in order for a human exchange to be accomplished—this has quite vanished. Dawn kisses the continents one after the other, and as it does, a series of coded communications hustles itself along the surface of the earth, relayed from point to point by fiber-optic cables, or bouncing in a triangle from the earth to synchronous satellites, serene in the cloudless sky. There is good news in Lisbon and bad news in Seoul, or the reverse; mountaineers reaching the summit of K2 send messages to their fearful spouses and then slip into sleep, laptops beeping until their batteries (and their owners) go dead; there is data everywhere, and information on every conceivable topic: the way in which raisins are made in the Sudan, the history of the late Sung dynasty, telephone numbers of dominatrices in Los Angeles, and pictures, too. A man may be whipped, scourged, and scoured without ever leaving cyberspace; he may satisfy his curiosity or his appetites, read widely in French literature, decline verbs in Sanskrit or scan an interlinear translation of the *Iliad,* discovering the Greek for "greave" or "grieve"; he may scout the waters off Cap Ferrat—somewhat gray with pollution as I recall—or see the spot where treasure lies buried in the wine-dark sea off the coast of Crete. He may arrange for his own cremation on the Internet or search out remedies for obscure diseases; he may make contact with covens in South Carolina, or exchange messages with chat groups who believe that Princess Diana was murdered on instructions tendered by the house of Windsor, the dark, demented old Queen herself sending the orders that sealed her fate. For all the great dreams profitlessly invested in the digital computer, it is nonetheless true that not since the framers of the American Constitution took seriously the idea that all men are created equal has an idea so transformed the material conditions of life, the expectations of the race.

INTRODUCTION

The Jeweler's Velvet

Two ideas lie gleaming on the jeweler's velvet. The first is the calculus, the second, the algorithm. The calculus and the rich body of mathematical analysis to which it gave rise made modern science possible; but it has been the algorithm that has made possible the modern world.

They are utterly different, these ideas. The calculus serves the imperial vision of mathematical physics. It is a vision in which the real elements of the world are revealed to be its elementary constituents: particles, forces, fields, or even a strange fused combination of space and time. Written in the language of mathematics, a single set of fearfully compressed laws describes their secret nature. The universe that emerges from this description is alien, indifferent to human desires.

The great era of mathematical physics is now over. The three-hundred-year effort to represent the material world in mathematical terms has exhausted itself. The understanding that it was to provide is infinitely closer than it was when Isaac

Newton wrote in the late seventeenth century, but it is still infinitely far away.

One man ages as another is born, and if time drives one idea from the field, it does so by welcoming another. The algorithm has come to occupy a central place in our imagination. It is the second great scientific idea of the West. There is no third.

An algorithm is an *effective procedure,* a way of getting something done in a finite number of discrete steps. Classical mathematics is, in part, the study of certain algorithms. In elementary algebra, for example, numbers are replaced by letters to achieve a certain degree of generality. The symbols are manipulated by means of firm, no-nonsense rules. The product of $(a + b)$ and $(a + b)$ is derived *first* by multiplying a by itself; *second,* by multiplying a by b twice; and *third,* by multiplying b by itself. The results are *then* added. The product is $a^2 + 2ab + b^2$ and that is the end of it. A machine could execute the appropriate steps. A machine *can* execute the appropriate steps. No art is involved. And none is needed.

In the wider world from which mathematics arises and to which the mathematician must like the rest of us return, an algorithm, speaking loosely, is a set of rules, a recipe, a prescription for action, a guide, a linked and controlled injunction, an adjuration, a code, an effort made to throw a complex verbal shawl over life's chattering chaos.

My *dear boy,* Lord Chesterfield begins, addressing his morganatic son, and there follows an extraordinary series of remarkably detailed letters, wise, witty, and occasionally tender, the homilies and exhortations given in English, French, Latin, and Greek. Dear boy is reminded to *wash* properly his teeth, to *clean* his linen, to *manage* his finances, and to *discipline* his temper; he needs to *cultivate* the social arts and to *acquire* the art of conversation and the elements of dance; he must, above all, *learn* to please. The graceful letters go on and on, the tone regretful if only because Lord Chesterfield must have known that he was volleying advice into an empty chamber, his son a dull, pimpled, rather loutish young man whose wish that his elegant father would for the love of God just stop talking throbs with dull persistence throughout his own obdurate silence.

The world the algorithm makes possible is retrograde in its nature to the world of mathematical physics. Its fundamental theoretical objects are *symbols,* and not muons, gluons, quarks, or space and time fused into a pliant knot. Algorithms are human artifacts. They belong to the world of memory and meaning, desire and design. The idea of an algorithm is as old as the dry humped hills, but it is also cunning, disguising itself in a thousand protean forms. With his commanding intelligence, the seventeenth-century philosopher and mathematician Gottfried Leibniz penetrated far into the future, seeing universal calculating machines and strange symbolic languages written in a universal script; but Leibniz was time's slave as well as her servant, unable to sharpen his most profound views, which like cities seen in dreams, rise up, hold their shape for a moment, and then vanish irretrievably.

Only in this century has the concept of an algorithm been coaxed completely into consciousness. The work was undertaken more than sixty years ago by a quartet of brilliant mathematical logicians: the subtle and enigmatic Kurt Gödel; Alonzo Church, stout as a cathedral and as imposing; Emil Post, entombed, like Morris Raphael Cohen, in New York's City College; and, of course, the odd and utterly original A. M. Turing, whose lost eyes seem to roam anxiously over the second half of the twentieth century.

Mathematicians have loved mathematics because, like the graces of which Sappho wrote, the subject has wrists like wild roses. If it is beauty that governs the mathematicians' souls, it is truth and certainty that remind them of their duty. At the end of the nineteenth century, mathematicians anxious about the foundations of their subject asked themselves why mathematics was true and whether it was certain and to their alarm discovered that they could not say and did not know. Working mathematicians continued to work at mathematics, of course, but they worked at what they did with the sense that some sinister figure was creeping up the staircase of events. A number of redemptive schemes were introduced. Some mathematicians such as Gottlob Frege and Bertrand Russell argued that mathematics was a form of logic and heir thus to its presumptive certainty; following David Hilbert, others argued that mathematics was a formal game played with formal symbols. Every scheme seemed to embody some portion of the truth, but no scheme embodied it all. Caught between the crisis and its various correctives, logicians were forced to organize a new world to rival the abstract, cunning, and continuous world of the physical sciences, their work transforming the familiar and intuitive but hopelessly unclear concept of an algorithm into one both formal and precise.

Their story is rich in the unexpected. Unlike Andrew Wiles, who spent years searching for a proof of Fermat's last theorem, the logicians did not set out to find the concept that they found. They were simply sensitive enough to see what they spotted. But what they spotted was not entirely what they sought. In the end, the agenda to which they committed themselves was not met. At the beginning of the new millennium, we still do not know why mathematics is true and whether it is certain. But we know what we do not know in an immeasurably richer way than we did. And learning this has been a remarkable achievement—among the greatest and least-known of the modern era.

In the logician's voice:

an algorithm is

a finite procedure,

written in a fixed symbolic vocabulary,

governed by precise instructions,

moving in discrete steps, 1, 2, 3, . . . ,

whose execution requires no insight, cleverness, intuition, intelligence, or perspicuity,

and that sooner or later comes to an end.

CHAPTER 1

The Marketplace of Schemes

Some philosophers see into themselves, and some into their times; still others forge an alliance with the future, scribbling their secrets late at night and speaking in whispers to the insubstantial and impatient souls that are gathered around their study door, dying to be born. Years go by and the dust of time collects. A new world is made. Things whir and pop and sizzle. There is the clatter of dropped dishes and laughter in the dark. Heels clack on the pavement. Taxis wheel and honk, and sanitation trucks bang down city streets. The sky is filled with electromagnetic pulses. The smoldering red sun edges over the horizon. Alarm clocks ring or chime or tinkle. Radios burst into chatter and still some silken thread of memory ties the present to the past so that, pausing for just a moment, we see ourselves reflected in a scholar's eyes, the calm smile in the portrait knowing and at peace.

The footman scratches discreetly at the library door. Gottfried *von* Leibniz, formerly Gottfried Leibniz, and his father before that Leibnütz, the *von* derived from god-knows-where, may now be observed entering the great audience room of history. In a gesture of practiced fluency, he bends slightly from the waist, his right foot obliquely in front of his left, and sweeps his arm in a half circle from his shoulder to his hip. His calm vitality rises like heat from a hot stove.

Leibniz was born in 1646, in Leipzig as it happens, on what is now the Leibnizstrasse, just off the Rosenthal, the valley of roses around which the city is built. His bustling entrance onto the scene occurred just two years before the Peace of Westphalia brought the Thirty Years' War to an end; and unlike so many of his countrymen, regularly the victims in their thirties or early forties of typhoid or smallpox or some ghastly festering infection, he died in his bed at a relatively ripe seventy years of age, spending his last hours discussing alchemy with his physician. The superb portrait by Andreas Scheits that hangs in the Uffizi in Florence shows him in court regalia, the rich brocaded silk of his shirt closed at the throat with a jeweled stud. His face is long but not narrow. The nose is majestic, creasing that face like a mountain range. The dark eyes are dignified, measured and reflective.

The portrait's calm is somewhat at odds, of course, with the exuberant chaos and disorganization of the man's life. During his time on the European stage, Leibniz traveled everywhere and saw everyone, crossing the continent again and again, listening to the gabble of his traveling companions (when he was not being conveyed in his personal coach), eating rustic tavern food, and staying overnight at rude, smoky roadside inns. As a young man, he lived for a time in Paris, where he made the acquaintance of the philosophers Antoine Arnauld and Nicolas Malebranche, and where he took instructions in mathematics from the Dutch physicist Christian Huygens. Presented with those dutiful Dutch hors d'oeuvres, Leibniz must have realized as he was dabbing his lips, that if he were to address a main course he would have to prepare it himself. As he did, establishing himself in a just a few years as a mathematician of uncommon insight and power. In 1673, Leibniz crossed the English Channel in a yacht in order to present himself to the Royal Society in London. He had let it be known that his beautifully made wood and brass calculating machine could do multiplication and division, as well as addition and subtraction. The English were interested but skeptical. The demonstration commenced, but at a crucial moment, Leibniz was observed carrying over remainders by hand.

Peter the Great greeted Leibniz in Russia, and he was at home everywhere in the German states and duchies, a close friend until her death of Princess Sophie of Prussia, instructing the princess in fond familiar letters of the particulars of his philosophic system. The scene is irresistible, Leibniz having written his letters with immense care, the dignified old woman scurrying away from court to read them in peace, trying to make sense of substances and monads, categories, contingencies, and the calculus, sifting the letters for their human warmth. Writing in French or in the German that he adored or in the supple and elegant Latin that he had taught himself at the age of eight, Leibniz kept up a correspondence with more than six hundred scholars and so maintained a vibrant presence in the marketplace. He knew everyone and everyone knew him, the organ of his interests booming over mathematics, philosophy and law, history and the design of hydraulic presses, silver mining, geography, political theory, diplomacy, windmill construction, horticulture, library organization, submarines, water pumps, clocks, and genealogy—this last, a chore imposed upon him by the house of Brunswick, its plump, gouty princes unaccountably interested in their weedy forebears. He wrote continually, but rarely for publication, and while his essays and epistles and night jottings are often jerky and incomplete, the genius that they reveal has the lucidity of gushing water.

Along with his rival, the secretive and suspicious Newton, Leibniz discovered (or invented) the calculus, the indispensable mathematical tool that made the scientific revolution possible, and he invented as well a brilliant and flexible notation for its fundamental operations, and so, much to Newton's indignation, casually dropped a glittering reference to himself into the stream of time. He discovered the correct equation for the catenary curve (among others), expressing in algebraic terms the sinuous way in which an iron cable suspended from two struts sinks and rises in a concave arc; he played brilliantly with infinite series, the jewels of analysis; and two hundred years before Henri Poincaré tackled the wet works and the masonry, he laid the foundations of mathematical topology, seeing (or sensing) that when quantitative measurements of distance and degree are withdrawn from geometry, what remains is a pure science of shape, a catalog of continuous deformations (as when the face of a Korean athlete is morphed into a flowering tree in those bizarre advertisements for running shoes).

Philosophers know Leibniz as the creator of a fantastic and elaborate metaphysical system, one in which the universe lies reflected in each of its monads, the scheme still of some interest and so saved from silliness because it seems forever to be trembling on the cusp of revelation without ever quite dropping into its cup. Like Kurt Gödel, the great logician of the twentieth century and his ectoplasmic kinsman, Leibniz was an optimist, his conviction that the world could be made no better at odds with the general opinion that it could be made no worse.

This is the Leibniz that we see in retrospect, a dynamic daytime figure, the largest and most ebullient Continental personality of his time, but unlike Newton, indeed, unlike anyone at all, Leibniz had an uncanny feel for the shadows behind the substance of his thoughts; his notebooks reveal a man dealing with problems he is barely able to describe, the play of his fine intelligence flicking between the seventeenth century and the far future. They reveal his obsessions, the topics to which he returned again and again as his mind stretched and unfolded itself. The idea of an algorithm takes shape in those notebooks, shaking off the dust of centuries as for the first time it moves into the courtyard of human consciousness.

THE LOGICIAN'S CLAMP

Curiously enough, it is logic that catches and then grabs Leibniz's attention, and I say "curiously enough" because by the seventeenth century, logic had become a discipline immured in custom and convenience, a barely noticed intellectual tool, its bones bleached white by time. Logic is the science of correct reasoning—*correct,* as in right, proper, indubitable, ineluctable, irrefragable, necessary, categorical; and *reasoning,* as in the passage from premise to conclusion, from what is assumed to what is inferred, or from what is given to what is shown. It was Aristotle who in the third century B.C. noted and then named the argumentative forms, codified inferences, and assigned the subject its characteristic shape—the work of civilization.

Aristotelian logic is categorical and so reflects the play between "all" and "some," the two terms of quantification giving rise to four statement forms:

All *A*s are *B*
No *A*s are *B*
Some *A*s are *B*
Some *A*s are not *B*.

The letters *A* and *B* function as symbols with a variable message, standing indifferently for any sensible subject: All *men* are *mortal*; no *whales* are *fish*; some *secrets* are *sinister*; some *signs* are not *seen*. Within Aristotelian logic, categorical statements are organized as *syllogisms,* arguments in which a conclusion follows from a major and a minor premise. All mammals are warm-blooded; but then again, all dogs are mammals. It follows that all dogs are warm-blooded. The drumbeat of three quantifiers sweeps this argument from the mammals to the dogs.

Although medieval schoolmen and Arab logicians fiddled with the system, Aristotelian logic does not move appreciably beyond the syllogism—it does not move beyond the syllogism at all—and so cannot describe the movement of the mind as it passes from the premise that a horse is an animal to the conclusion that the head of a horse is the head of an animal. This is an inference that cannot be shoehorned into a syllogism, no matter the shoehorn, no matter the syllogism.

But for all of its obvious limitations, the hold of Aristotelian logic on succeeding generations was nonetheless powerful and sustained. The brilliant and unique culture of the ancient Greeks exhausted itself in the still splendid sunlight. Barbarians came to roam the torn and tattered margins of the Roman Empire. A sophisticated Christian culture rose in Europe, far from the Mediterranean Sea. Anonymous architects and builders constructed astonishing cathedrals in fields where wildflowers grew. The medieval world disappeared around a bend in the river of time, replaced at last by a culture recognizably contiguous to our own, and through it all, the sun having risen on bright bursting warlike cities and set on ruined châteaus perched on lonely hills of pine and juniper, Aristotle's logic remained the sign and the standard, men and women who spoke in alien jabbering tongues endeavoring to organize their thoughts into major and minor premises, worrying about the distribution of terms, the solemn ancient forms dominating their imagination.

THE MINEFIELD OF MEMORY

■ ■ ■ ■ Friends and students now collect themselves from various law firms and trauma centers and arbitrage houses to spread themselves in front of me. I am glad to see them. They are still young and so am I. We are trotting together over the minefield of memory. "Bowser cannot fly," I say with brisk conviction, "because *really dogs really cannot really fly*." Whatever the dogs may wish for (wings, cloud-capped fire hydrants, a nibble of celestial kibble), the truth is a tough taskmaster. Her dark brown eyes still flashing, *Jacqueline Hacquemeister*, LL.B. (and formerly *Jackie*), rises from her seat and then at once subsides: she is far too intelligent to be taken in and by now she has acquired poise. The issue is settled. Bowser cannot fly.

But the logician, I go on to say, quite without saying anything aloud at all (and so communicating in the way I always wished I could), *is* prepared to accept worlds in which dogs can fly and pigs can speak and women like the flowers in their hair remain forever young.

Ms. Hacquemeister snorts the pharyngeal snort she used to snort, low, sharp, devastating, but oddly disconcerting nonetheless.

No, really. The inference proceeds hypothetically. *If* all dogs can fly—and there follows a cascade of canine consequences. In all of this, truth plays an ancillary role. An argument is assessed by the validity of its inference and not the truth of its premises. Within the ambit of a valid argument, *if* the premises are true, the conclusion *must* be true as well, the logician's clamp fixing propositions in an indissoluble matrix.

Ronald Kemmerling, M.D., looks up at me with eyes strained from lack of sleep. I remember that he used to sleep very soundly in class, his head dropping to his chest and then jerking upward. (He now directs the Nathan P. Trauberman

Memorial Transplant Center at the Mount Christopher Medical Center in Yonkers, New York. God punishes every man according to a special plan.)

"Flying dogs?" he mutters, letting his head droop for old time's sake. I am talking hypothetically, Doctor. And I don't mind if you nap. I may need you more than you need me. But look, dogs really aren't the point. Logic is a formal discipline. Dogs enter the picture as an instance. As they are so often in real life, they are simply along for the ride.

The logician's clamp tightens itself over propositions, but we feel the clamp and are bound by its power because inference is a mental motion, one that proceeds by means of the soft furry pop of intuition. Witness thus Bowser's aeronautical promotion: *All dogs can fly*. Pop. *Bowser is a pooch*. Pop. Hey! Look at that—*the damn dog is flying*. Except for the fact that Bowser *is* a pooch, these little explosions are counterfactual. It does not matter. The logician's clamp has tightened itself, the flow of inference matched only by the flush of infatuation as human activities falling under the control of the inexorable.

Ms. Hacquemeister beams suddenly, splitting the air with her lovely smile; I remember that she once brought her dog to class, an evil-smelling, friendly Labrador.

Arnau de la Riviere (formerly Arnie Kahane and the man who at a class party held at a French restaurant ordered *bifteck tartare très saignant*) smooths the sleek sides of his gray hair backward, as if to remind me that he is due shortly at the French embassy, where he represents American commercial interests in perfect French.

But beyond saying that it has tightened itself, that clamp, I am hard pressed to say *why*. And while everyone else in class can see the inference, or feel it moving through the muscles of their mind, no one can explain what he or she sees; and no one—especially not Jacqueline Hacquemeister, who has argued before the justices of the Supreme Court—is quite prepared to repose his or her confidence in what is purely a feeling, an impression of rightness.

Friends and students shuffle in impatience, their hands inching toward their cellular telephones. They are late. And so am I. Class dismissed.

■ ■ ■ ■

THE VIBRATING STRING OF DISCONTENT

Alone among his great contemporaries, Leibniz placed a plump index finger on this oddly vibrating string of discontent. And alone among his great contemporaries, he saw—in outline, at least—the structure of a system in which the mental motion involved in inference might be explained *and* ratified by a simple—a mechanical—procedure. A universal affirmation says that all *A*s are *B*, the dogs reposing among the mammals, the poets among the writers, the trout among the fish. But the fact conveyed by saying that all *A*s are *B* may be conveyed as well by saying that *both A and B are A*.

This is not obvious, but it is so. If all dogs are mammals, then the only creatures that are both dogs and mammals are just the dogs.

The fact that all *A*s are *B* may be conveyed by saying that both *A* and *B* are *A* suggests a novel possibility of representation, with all *A*s are *B* expressed as the algebraic identity $A = AB$. Here *AB* designates items in both *A* and *B*, what later-day logicians would call the intersection of two sets (the dogs and the mammals).

The algebraic invigoration of a categorical proposition now makes for an algebraic interpretation of the categorical syllogism:

All As are $B \rightarrow A = AB$
All Bs are $C \rightarrow B = BC$
All As are $C \rightarrow A = AC$

And it is algebraic invigoration that offers the entirely unexpected prospect of subordinating inference to a checklist, one contingent only upon operations defined over *symbols*.

In what follows, the categorical syllogism is on the left, the algebraic version on the right (parenthetical remarks offer an explanation for the transformation):

All dogs are mammals	1. A = AB	Check (this is what we are given)
All mammals are animals	2. B = BC	Check (this is given, too)
	3. A = ABC	Check (BC has been substituted for B in line 1)
All dogs are animals	4. A = AC	Check (A has been substituted for AB in line 3)

Within the categorical syllogism, ordinary language represents the ordinary flow of inference. Two premises are given; there is a plash of insight, and one step undertaken. The mind hops right along, not quite knowing where it is going but getting there nonetheless. On the right, a checklist does its work. The logician's clamp retains its force of old, but the inferential steps involve no more than the substitution of symbols for symbols, with the anchor of inference embedded in identities. Inference now proceeds from one identity to the next; no plash of insight is involved, only the solid satisfying ratcheting sound of symbols being substituted for symbols.

Whatever its merits as a scheme of inference, that checklist has by now achieved an oddly troubling form of generic familiarity. *If you wish to find out your current balance, press* *one; if you have earned less than $23,000 in the last fiscal year but more than $14,000, then add lines two and three; if you suffer from night sweats and the most peculiar itching . . . ; if this is Ralph calling, hang up and die . . . Mom, if it's you, I'm at the hairdresser's . . . if this is Bob, I'm so glad you called, so very glad.*

LET US CALCULATE

■ ■ ■ ■ John Frederick, the duke of Brunswick, suffers from piles and from rheumatism and from the chilblains. A sore on the inside of his cheek has refused to heal. There are angry red cracks on the skin between his toes, and when the night wind rattles the mullioned windows of his bedroom, he finds himself unable to sleep for more than two hours before the shrill urgency of his bladder forces him to turn his attention to the porcelain bedpan.

He is seated at his enormous desk, its top littered with sheets of vellum, ledgerbooks, scraps of paper, rolled-up maps, a treatise on hydrology, inkwells, and an ornate wooden box containing a collection of quills.

As Gottfried Leibniz seats himself opposite the desk, John Frederick lifts his pear-shaped head, the cheeks sagging into two pouches, and with a mute feeling of dismay, recalls that he has agreed to receive his counselor and court historian. "My dear fellow," he says graciously. The grandfather clock standing at the library's west wall solemnly sounds the hour.

Leibniz clears his throat, delicately raising his wrist to his lips. He has been given fifteen minutes to explain his new system.

"Not long ago, Your Excellency," he says, "some distinguished person devised a certain language or Universal Characteristic in which all notions and things are nicely ordered, a language with whose help different nations can communicate their thoughts, and each, in its own language, read what the other wrote."

The duke wrinkles the thick skin of his brow, a gesture that makes it seem as if his plump cheeks are being drawn by purse strings; he reflects that he has never had the slightest difficulty communicating his thoughts so long as people understand German or speak a proper French. A certain familiar fullness in his bladder reminds the duke that he must soon relieve himself.

"But no one," Leibniz continues, "has put forward a language or Universal Characteristic which embodies, at the same time, both the art of discovery and the art of judgment."

As Leibniz pauses and looks up, the duke casts about for a response. He can think of absolutely nothing to say.

"If we had it," Leibniz goes on, his voice low, urgent but curiously toneless, "we would be able to reason in metaphysics or morals in much the same way as in geometry and analysis. If controversies were to arise, there would be no more need of disputation between two philosophers than between two accountants. For it would suffice to take their pencils in hand, sit down to their slates, and to say to each other (with a friend as witness, if they liked): Let us calculate."

The duke of Brunswick, his issue destined for the British throne but his attention fatally diverted by his screaming bladder, waggles his pink palms, rises rapidly from his seat, and without a word scuttles down the long library corridor to the massive wooden double doors, where, after pulling at the red tasseled bell rope, he waits in an agony of impatience for a footman to appear.

Unperturbed, Leibniz continues to sit as he has sat, his foot properly extended, his buttocks perched decorously on the edge of the red satin chaise.

■ ■ ■ ■

A SCHEME

Beyond what he has said, Leibniz has a scheme, of course, if only because everyone in the seventeenth century has a scheme, the things popping up like mushrooms after a heavy rain, and it is only bad luck that has prevented him from asking his patron for funds to put his scheme into practice. Like almost everyone else, Leibniz thinks of his scheme in terms of an encyclopedia, his omnivorous intelligence demanding at every turn that the full range of human knowledge be spread out on the printed page, compiled in books, massed in libraries, collected in institutions. But while ordinary men were interested in ordinary encyclopedias, Leibniz was interested in something else and something more, an encyclopedia of human concepts, an alphabet of human thoughts—*humanity, revenge, piety, beauty, happiness, goodness, pleasure, truth, greed, hygiene, procedure, rationality, decorum, alacrity, motion, peevishness, ductility, manners, justice, politesse, competence, war, art, computation, duty, work, language, rubbish, information, feminine, fairness*—one comprehensively containing *every* human concept and so *every* human thought.

The idea of a great, a *complete,* alphabet of human thought suggests that in this regard, as in so many others, Leibniz was striking for the fundamentals; however he may have executed his idea—in truth, he got no farther with his idea than the idea itself—something like the alphabet must exist if the mind is ever to be an organ accessible to itself. His grand if unfulfilled vision rests on two superb and controlling insights. However many complex concepts there may be, the number of simple human concepts must be finite (they come to an end, those simple concepts) and what is more, they are discrete (unlike heat haze or moods or the moon seen through clouds, they have natural boundaries).

If there are only finitely many simple concepts, some principle of organization must be at work in thought, one orchestrating the way in which concepts are combined. Otherwise ordinary speech would be simply a matter of blurting out the names of elementary concepts, language nothing more than a series of verbal tags.

Physical objects suggest an obvious principle. Wholes have parts, the doughnut resolving itself into particles of flour and butter, the building into bricks, and the book into its pages. It is precisely the relationship of part to whole that Leibniz saw as coordinating the construction of concepts. *Concepts* now, and not things. Vichyssoise is a cold soup. Being cold and being a soup are just the constituents involved in being vichyssoise. In turn, being cold goes down to its parts, and ditto for being soup. Those parts go down farther, the dissection of concepts proceeding until the dissector has reached concepts that can no longer be dissected into parts because, like electrons in particle physics, they are absolutely simple.

Ultimately, Leibniz argued, there are only two absolutely simple concepts, God and Nothingness. From these, all other concepts may be constructed, the world, and everything within it, arising from some primordial argument between the deity and nothing whatsoever. And then, by some inscrutable incandescent insight, Leibniz came to see that what is crucial in what he had written is the *alternation* between God and Nothingness. And for this, the numbers 0 and 1 suffice.

Twinkies and diet Coke in hand, computer programmers may now be observed pausing thoughtfully at their consoles.

CHECKLIST

■ ■ ■ ■ We live amidst the infernal clutter of things (me, especially, it seems, and just where *are* my cigarettes?), but we make our judgments about facts. A bowl of vichyssoise rests among the things, but the judgment that vichyssoise is cold among the facts. So, too, the judgment that candy is sweet, that water boils at 212 degrees, and that the pharaohs departed for eternity wrapped in three hundred yards of treated linen.

Validity is the touchstone of inference, and truth of judgment: the fact that vichyssoise is cold ratifies the judgment that vichyssoise is, indeed, cold, and the judgment that vichyssoise is cold expresses the fact that vichyssoise is cold.

These very modest observations, which have been known to embroil philosophy departments in dispute for decades, suggest what judgments do, but not what they are. Leibniz is persuaded that in knowing what judgments do, *he* knows what judgments are, his assessment of judgment, interestingly enough, entirely consistent with his assessment of inference. "Concepts," he reminds me, having sat himself by my desk late one night (his lush old-fashioned wig proving irresistible to my cats, who have come creeping from their tower to bat at it), "have parts." Judgment is nothing more than an act of revelation showing that, like one of those dowdy Russian dolls reposing within yet another dowdy Russian doll, one concept is included within another. To say that vichyssoise is cold is to say only that the concept of being cold is a part of the concept of being vichyssoise.

The analysis now proceeds by means of a stream of impulses. First, the encyclopedia of human concepts is brought into service (some centuries before its completion), Leibniz thumbing through the entries for my benefit. The dissection of "vichyssoise" into its parts having already been completed (this is an *assumption*), Leibniz needs only to flip through the *V*s—"vapid," "various," "velocity," "venial," and so to "vichyssoise"—in order to see its conceptual constituents.

"Here it is," he comments with satisfaction in beautifully

modulated English, his skill with languages having always stood him in good stead.

The encyclopedia's entry is brevity itself. (I am transcribing things by means of a forward-looking form of memory):

> *Velocity*...
> *Vichyssoise:* cold, soupy
> *Victim*...

In a later edition of the encyclopedia (the 23rd), the declaration is compressed still farther:

> *V(elocity)*...
> *V:* C + S
> *V(ictim)*...

The addition sign gives notice that the encyclopedia is *not* a dictionary, one in which *vichyssoise* is defined as a cold soup. Entries in the encyclopedia are *lists* signifying that *these* concepts are the parts of *that* concept. These serve precisely the same role in the economy of judgment as lists identifying the constituents of breakfast cereals. They tell the reader (of the encyclopedia) or the consumer (of all that gloop) just what the concept or the cereal is made of.

Judgment now proceeds by means of yet another checklist:

1. *Consider:* "Vichyssoise is cold"	Check
2. *Look up:* "vichyssoise"	Check
3. *List* entries: "vichyssoise"	Check
4. *If* "cold" is an entry *then*	
5. *Accept:* "Vichyssoise is cold"	Check
6. *If* not, *then*	
7. *Reject:* "Vichyssoise is cold"	Check

Let me rehearse this somewhat disorderly scene. The soup arrives. Vichyssoise. And *lo*, the stuff is cold. Without pause or panic, the mind proceeds from what the palate perceives. Vichyssoise *is* cold. Introspection yields only a mysterious blue blur, self-inspection self-defeating in this case as in so many others. Now step back: ignore the blur. Judgments may be depicted at a distance by means of a play of symbols, with the fact that vichyssoise is cold both explained and ratified—just what the inferential checklist does, recall—by the fact that being cold is a part of being vichyssoise. Whatever the *mind* may be doing—scratching itself, worrying about the onset of Alzheimer's, getting ready for a nap—the checklist is simple, straightforward, and objective, resting on nothing more subtle than the principle animating those primly proliferating Russian dolls.

Sometime later, I think to ask Leibniz for the source of his confidence in his own scheme. Without a word, he repairs again to the encyclopedia, resting his finger on the third line, which he proceeds to tap:

> *jagged* . . .
> *jealousy* . . .
> *judgment:* part, whole
> *justice* . . .

I am for a moment (but only for a moment) at a loss for words. Then I recover myself, sputtering: "All that you've shown is that this scheme of yours justifies itself."

Leibniz smiles an ineffable smile. "In the end," he says, "every scheme and every science is justified by itself or it is not justified at all."

Against every expectation, Leibniz envisioned his encyclopedia as a satisfying practical device, a bridge of brotherhood; he imagined that with the encyclopedia complete, every human concept would be assigned a symbol, an odd kind of glyph, so that looking at the symbol ✸ one could see at a glance that it designated a snowflake. (Certain languages such as Chinese are organized by means of this principle.) With the glyphs of this system in place, all forms of human communication become entirely mechanical.

It remains only to imagine the scheme in action. Departing, Leibniz turns to wave: ❂.

Understanding at once, I nod: ⊖.

He smiles sadly: ⋟.

I sign: ✢.

And then, as the light draws close, he turns and shuffles down the corridor of time, dragging his right foot slightly because he suffers so from gout.

■ ■ ■ ■

CHECK OFF

At the end of the twentieth century, Isaac Newton and Gottfried Leibniz may be discerned sitting in the cockpit of change, both men endeavoring strenuously to remove one another's hands from the controls. The origins of Newton's theory of gravity has become a myth, one in which the now-venerable combination of tree, apple, and bonk prompted Newton to the incredible and far-ranging hypothesis that a single force controls the behavior of objects in motion, from the place where the Magellanic Clouds gather in the weak light of space to the very surface of the earth, where apples, stock markets, bosoms, arches, and wisteria sooner or later droop or drop. Newton provided an explanation for gravity in terms of a law of universal attraction, one binding every material particle in the universe, the law expressing mathematically the attraction of matter for matter, but Newton's great chilly vision of what in the *Principia* he called "the system of the world" goes well beyond mechanics, his narrow obsidian eyes searching for a scheme by which every aspect of the physical world might be explained in terms of a complete and consistent set of mathematical laws. The scheme remains beyond our grasp, but not the search: contemporary physicists, like Newton before them, trot restlessly over the very same conceptual landscape, baying at dark shadows in the night.

If Newton proposed to construct a system of the world, Leibniz proposed instead—but there is no *instead,* you see. The two men's interests are simply not commensurable.

In comparison with the superbly focussed Newton, Leibniz seems at times to have been pursuing an agenda he could not completely articulate; no small wonder that his ideas skip the intervening centuries to topple somewhere beyond. If there is no grand scheme at work (at least in logic), there is nonetheless a progression in his thought, one in the direction of ever-increasing abstraction, so that in the end, like a Japanese painter eliminating every color from his palette except black (and trusting the paper itself to supply a contrastive white), Leibniz purged from his encyclopedia all traces of its content and attended only to what is left behind, a system of pure symbols and forms. "Combinatorics," he wrote, "treats of calculus in general, or of general signs or characters (such as *A, B, C,* where any one could be taken for another at will), and of the various laws of arrangements and transitions, or of formulas in general."

With these words, a different organization of experience creeps forward to announce itself. The material world recedes: *symbols* take center stage and are in command. Meditating on the meaning of certain mental motions that rumble far beneath the near-margin of consciousness, Leibniz endeavored to provide an account of inference and judgment involving the mechanical play of symbols and very little else. The checklists that result are the first of humanity's intellectual artifacts. They express, they explain, and so they ratify a power of the mind.

And, of course, they are artifacts in the process of becoming algorithms.

He suffered from gout and from colitis and from a desolate conviction that he had outlived his time, with even his luxuriant wig and elaborate clothing out of place at the German court. He retreated into the spaciousness of his thoughts. At last the noose of life grew tight. Leibniz withdrew to his bed. He refused the professional attention of his physician, knowing that the man had nothing better than a course of bloodletting or a session with leeches to offer. His mind was lucid until the end. Turning his face toward the wall as death approached, he drew the tasseled cap he wore in bed over his eyes, a ragged breath catching at his throat, and then lay silent.

Arriving late one night in northern Norway, I happened in my hotel room on one of those superb BBC science programs in which scientists expatiate on their own work. A number of cosmologists were discussing the origins of the universe. It is what they always discuss and the discussions are always fascinating, if inconclusive. It is Isaac Newton who dominates the various exchanges, his deep, powerful compelling intelligence bending every will to his own. And then a curious remark. "There are these recipes," one cosmologist said, "these equations that seem to rule the world."

These recipes? These equations? These *algorithms*?

CHAPTER 2

Under the Eye of Doubt

The seventeenth century is past, Leibniz and Newton having gone to where great men go; the eighteenth century, that, too, has disappeared, various lace-wrapped but lice-infested heads having toppled into various revolutionary baskets with various dreadful red wet plops. The Congress of Vienna, the revolutions of 1848, and the American Civil War swim into view and disappear. A word of praise for laudanum and tinctures of opiate, the stuff swallowed gratefully by women aged in childbirth and forever stuck in places where the prairie winds blow late at night. New governments in France and Germany. The nineteenth century now rounds itself, like a great clipper ship bending to the wind. At an international congress in Lausanne held in 1881, a group of bearded physicists gather for a formal portrait. They are wearing stiff black gabardine frock coats and staring into the camera with an expression of daffy confidence. Sunny skies prevail; there is everywhere the smell of lemons.

We know now that there were cold gray clouds scheduled to come skittering over those sunny skies, the clouds destined to drop their loads over science as well as life. In 1887, the American physicists Albert Michelson and Edward Morley observed experimentally that the speed of light seemed unaffected by its passage through the luminiferous ether, a conclusion at odds with Newtonian physics and with common sense, and, indeed, a conclusion at odds with the existence of the luminiferous ether. Eighteen years later, Albert Einstein resolved the contradiction in favor of the facts, arguing brilliantly that space and time were themselves relative, time slowing and space contracting as objects increased their speed. In the same year, he argued that light travels in quantum packets, like freight cars on a railroad train. Thereafter physics trudges off into its dark defile, at odds with common sense at every step. But there are clouds and then again there are clouds, and the ones that carried the hardest rain happened to cover mathematics and not anything else.

Mathematics?

ENTER GIUSEPPE PEANO

The essentials now. A hard walnut face, small black shrewd watchful staring peasant eyes, a lean long nose erupting into a tuber, his chin covered improbably with Lenin's scraggly beard, almost as if the famous figures on the European scene were in the habit of loaning one another facial hair. Peano was born in 1858, near the village of Spinetta, which lies within Italy's Cuneo province. This is the Piedmonte. There is everywhere in this countryside traces of the old, the Roman way of life, the open air reverberating with the rumble of imperial legions tramping through the fields before ascending the mountain passes toward Transalpine Gaul. They grow wheat and rice in the lowlands, and produce butter, cheese, and milk in the sub-Alpine meadows. Like the tidelands of South Carolina, this is land that is rich without being beautiful, the great open dun-colored fields crisscrossed with irrigation canals, the brown sun high in the smoky sky.

Peano was born the son of peasants, his family long rooted to the umber hills, and had he been born one hundred years earlier, he could have expected only that in time he might have replaced his father, marrying a stout-hipped local girl from the villages, living a life of quiet rural prosperity, the goats bleating on the hillside and the stone farmhouse smelling of bread and thick stews by midafternoon. But Peano was the beneficiary of an extraordinary educational system, one that offered the Italian peasant class opportunities for social promotion along a narrow but nonetheless smooth-paved corridor of professional attainment. His eldest brother became a land surveyor, the prosperous father of seven children; another brother became a priest, then as now a matchless sinecure in Italy; a sister married well, disappearing into domestic oblivion, children hanging on her apron strings and waving shyly; only his younger brother returned home from the village school to walk behind the farm's moody mules and work the fragrant fields.

As an urchin, Peano passed through the village school, and then as a stripling through the Liceo Cavour, and thereafter as a croaking adolescent through the Collegio delle Provincie, a branch of the University of Turin, one of those institutions established by the Italian state specifically to assist talented students from the provinces. There were classes five days a week, almost all in the sciences and mathematics, competitions, tests, stern oral examinations, these followed by an honorable mention after his first year. An *honorable* mention? Examinations again, this time in purely mathematical subjects, and thereafter, Peano, ensconced now at the University of Turin itself, his professional association there continuing until his death in 1932, is a university assistant, an extraordinary professor, a full professor, a Considerable Figure, one known to other Considerable Figures, the steps passing one after the other until they reveal the unmistakable emergence of an Historical Eminence, the lean feral face addressing students in class, mathematicians at international conferences, and the future at night, his voice curiously hoarse, the great man speaking with an endearing lallation for his entire life, not quite able completely to enunciate the Italian *r*, pronouncing it instead as the liquid Italian *l*.

RHAPSODY IN NUMBERS

The experiences of mathematicians are sometimes rhapsodic and therefore always tragic. Mathematicians age rapidly, reaching mathematical maturity before they have learned to look directly into a woman's eyes and declining thereafter with terrible and remorseless speed. The Fields Medal (the mathematical equivalent of a Nobel Prize) is awarded only to mathematicians under forty; at an age when surgeons-in-training are still learning the precise anatomy of the iliac crest, the great mathematicians have already felt cold winds sweeping down from the Aral Sea.

For the most part, it is true, ordinary men and women regard mathematics with energetic distaste, counting its concepts as rhapsodic as cauliflower. This is a mistake—there is no other word. Where else can the restless human mind find means to tie the infinite in a finite bow? Sometime in the seventeenth century, for example, mathematicians discovered that addition could be extended to infinite sums. By an infinite sum, I mean an infinite sum. The numbers go on and on, and yet somehow reach a sum somewhere:

$$a_1 + a_2 + \ldots + a_n + \ldots = \sum_{i=1}^{\infty} a_i.$$

Lowercase italic letters in this expression stand for numbers; subscripts serve to tag the numbers in a sequence (the first number, the second number, and so on up), and the large Greek sigma indicates addition carried out *from* an initial number *to* infinity.

Concept and the notation are at work in the following series:

$$1 + \frac{1}{2} + \frac{1}{4} + \frac{1}{8} + \cdots + \frac{1}{2^{n-1}} = \sum_{i=1}^{\infty} a_i,$$

the symbols encouraging the mathematician to *take* the numbers as they are written, *imagine* them going on forever in just that peculiar way, the denominator of each fraction determined entirely by its place in the sequence, (so that $1/8$ is just $1/2^{n-1}$, which in turn is $1/2^3$, which in turn is $1/8$), and then *figure out* their sum.

These requests—*take*, *imagine*, *figure out*—are expressed in ordinary English and ask that something be done: so much is clear; but contemplating addition carried out forever, the mind suddenly skates over ice where previously there was a concrete walkway, whooshing forward without pause or purchase.

Infinite addition requires the domestication of the infinite, the requisite effect achieved by means of two mathematical crampons. The first involves the segregation of series into finite partial sums (finite, partial, and so *ordinary*); the second, the concept of a limit.

Given that

$$1 = S_1,$$

$$1 + \frac{1}{2} = S_2,$$

$$1 + \frac{1}{2} + \frac{1}{4} = S_3,$$

the sum of the series as an infinite whole is defined by asking whether $S_1, S_2, S_3, \ldots$ are tending toward any particular number as a limit. If so, that is the sum; if not, not. Partial sums and limits suffice to bring the infinite to heel—at the boot of the number 2, in this case.

There is in this no hint of the rhapsodic, but infinite series are very much like jewels, and only a slight rotation of the subject in one's palm reveals a marvelously glinting face.

The series

$$1 + \frac{1}{2^2} + \frac{1}{3^2} + \frac{1}{4^2} + \cdots$$

looks for all the world as if it might converge to some familiar number; but Leibniz could not determine its sum, and neither could Jacques Bernoulli (and neither could I, for that matter), the question remaining open until Leonhard Euler traced those glinting rays of light back to their face and discovered the partial sums converging to $\pi^2/6$.

The number π represents the ratio of the circumference of a circle to its diameter and so embodies a mathematical constant, one whose value lies just north of the number 3. The series is itself composed of simple-seeming fractions. And yet these numbers converge to the ratio of π^2 to 6, thus revealing a glittering connection between geometry and arithmetic, the connection all the more glittering because all the more arbitrary. Why π? Why π^2? Why 6? Why the ratio of π^2 to 6?

Why, indeed.

I hear myself asking these questions in class and then late at night, and I ask them because I do not know the answers, not then, not now. All that the mathematician in me can see is the jewel.

AT THE LIMIT

I must now allow a little rain to fall on the flames of my own prose. Whatever the time spent swooning, the mathematician, like the lepidopterist, is *professionally* engaged in an effort to limn the loveliness that he sees, the mathematician fixing in formalism what the lepidopterist fixes in formaldehyde. But the desire to see and the desire to ratify what one has seen are desires at odds with one another, if only because they proceed from separate places in the imagination. A few first drops have just fallen.

After centuries in which the jewel of mathematics emitted only pale fires, mathematics burst into light in the seventeenth century with the discovery of the calculus and thereafter continued to flash furiously as mathematical analysis, algebra, arithmetic, number theory, and non-Euclidean geometries came into existence or achieved a sudden and disturbing maturity. A universe was created from that flash in the seventeenth, eighteenth, and nineteenth centuries, replete with cosmic dust, winking distant stars, strange galaxies, and planets that look much like our own.

But with the intellectual ecstasy engendered by creation came intellectual dread as well, the sense, quickening as the decades stood, shook themselves, and stamped off the field of history, that no one quite knew why mathematics was true and whether it was certain. The concepts needed to express the new body of mathematics were often incoherent. The calculus, philosophers observed (with no small satisfaction), reduced itself to absurdity by an invocation of infinitesimals, numbers smaller than any other numbers but not zero. The carefully crafted definitions introduced for prophylactic purposes proved alarmingly complex. I have written of limits in terms of tendings, and for amateurs (like us), that suffices. It sufficed as well for eighteenth-century mathematicians. It did not suffice thereafter. Mathematicians now compute limits—in class and in life—by appealing to the following definition:

> *A series (those partial sums, for example)* a_1, $a_2, \ldots, a_n, \ldots$ *converges to a limit* L, *if, for every positive number* ϵ, *there is a value of* n *(which in*

> *turn depends on* $\in$*) such that for* a_n*, and for all numbers farther up the chain, the distance between* a_n *and L is less than* $\in$*.*

The number 2 is the limit of the series

$$1 + \frac{1}{2} + \frac{1}{4} + \frac{1}{8} + \cdots + \frac{1}{2^{n-1}} = \sum_{i=1}^{\infty} a_i,$$

because no matter how small the number $\in$ (say that it is 1/269,000), there is some number a_n in the sequence, such that the distance between a_n and 2 is less than 1/269,000 and stays less than 1/269,000 no matter how large n becomes.

This is hardly the stuff of intuition, and intuition is hardly prepared to deal with it (as many a twentieth-century student will attest). Nineteenth-century mathematicians discovered to their discomfort that as the conceptual machinery of mathematics became more precise, it became more difficult. Definitions that had plashed their way from textbook to textbook were discovered to be flawed, and flaws were detected in famous proofs, the radiant expectations of a professor in Basel dashed because a professor in Kiel discovered that the professor in Basel had in his celebrated demonstration assumed that certain functions could reach around their heads to scratch their right ears with their right hands when in fact they could get only so far as their noses.

The confusions were hardly restricted to the calculus. No one quite escaped falling under the eye of doubt. The story of Évariste Galois has by now entered mathematical and romantic myth. Born in 1811 in the French provinces, Galois was a mathematical rhapsodist, his mind, even in childhood, entirely unclouded, so that like so many great mathematicians, he seemed to catch the colors of mathematics on the wing. Unruly and intemperate, Galois antagonized his examiners at the École Polytechnique. His great work on group theory was accomplished before he was twenty, on the evening of his death, in fact, for he died in some idiotic duel over a woman the succeeding dawn, taking a bullet in his stomach and expiring in agony some hours later. At the moment his genius flowered and was then extinguished, the mathematicians of France, while aware of his powers, could not determine—they could not *decide*—whether his early work made any sense. "His argument," members of the Academy of Sciences wrote, "is neither sufficiently developed nor sufficiently clear to allow us to judge its rigor."

Its *rigor*. Meaning whether it was true. Meaning, God help us, *we* cannot tell and *we* do not know.

A HARD RAIN FALLING

Giuseppe Peano was awarded his *dottore* in mathematics in 1880, and for all the ceremonial rigmarole of his investiture into university life—brocaded gowns, tassels, mortarboards, strange caps, and Latin oaths—the profession of mathematics had by then existed for little more than two hundred years. Isaac Newton did attend classes given by the Lucasian professors of mathematics, Isaac Barrow, in mid-seventeenth-century Cambridge, but whatever Barrow knew of mathematics could easily have been inscribed in a small chapbook—half a hundred formulas, the axioms and theorems of Euclidean geometry, the rudiments of algebra (which Newton magnificently dismissed as "the analysis of bunglers"), something of the new Cartesian system of algebraic geometry, and beyond this, a jumble of confused and imprecise ideas about rigor, definition, and proof.

Beyond that, nothing.

Yet by the time that Peano mounted the worn wooden steps toward the lecturer's podium, the amateurs of hot genius, who had written with quill pens and perfumed their wigs, had done their work. Mathematics had changed from a subject quivering in a thousand living letters into something mummified and monumental, entombed within the Royal Society or various French academies and then entombed again within the great European universities. Peano is a *professore*; there have been *professores* before him. And there are questions that the *professores* ask that the amateurs of hot genius do not. How does one draw conclusions about fearfully abstract and subtle concepts so that the conveyance itself has the character of conviction? It is not faith that is wanted; neither is it insight or intuition.

It is certainty itself, the mathematician, like the lover, needing more to be sure than to be happy.

Rain is now general.

THE PODIUM OF HISTORY'S GREAT LECTURE HALL

Peano began his career as a *professore* of the infinitesimal calculus, and his contributions to the discipline were elegant, influential, and important. He constructed a proof for the uniqueness and existence theorems for first-order differential equations, one ratifying mathematical expectations that when it comes to these crucial instruments of description and discovery, solutions do exist and they are unique. But Peano was troubled from the first by the eye of doubt, the oily thing staring at him from underneath hooded lids and bushy brows, a gnawing and uneasy feeling that the foundations of mathematics were somehow infected; and in that odd way that great figures in the history of thought have of addressing one another across time, he seems disposed to have taken counsel from Gottfried Leibniz, the elegant rubicund courtier, with his massive wig and noble nose, whispering into the ear of the tense Italian mathematician, newly a *dottore*.

If there is any part of mathematics that would appear safe from scruple, it is arithmetic, the subject as familiar as childhood itself, the only part of mathematics lying resident in memory (along with telephone numbers, passwords, and the smell of roses in the rain); and, in truth, no one in his or her right mind would question the ordinary arithmetical exchange in which two mathematicians and two mathematicians combine to make four professors. Our certainty on this score is absolute, no matter how penetrating the eye of doubt. But like so many unwholesome strangers, doubt enters on the second floor, avoiding the living room, where there are bright lights and chamber music. Arithmetic is not simply a series of arithmetical exchanges or childhood maxims; the natural numbers, 1, 2, 3, . . . , go on forever, and so our confidence in arithmetic *as a system* goes beyond—*way* beyond—the confidence that we may repose in any of its finite parts.

The world of shapes, lines, curves, and solids is as varied as the world of numbers, and it is only our long-satisfied possession of Euclidean geometry that offers us the impression, or the illusion, that it has, that world, already been encompassed in a manageable intellectual structure. The lineaments of that structure are well known: as in the rest of life, something is given and something is gotten; but the logic behind those lineaments is apt to pass unnoticed, and it is the logic that controls the system.

Euclidean geometry proceeds from a finite set of axioms; from these, the mathematician derives various geometrical conclusions or theorems. The interior angles of a triangle sum to 180 degrees. The surveyor knows this by measurement, and knows what he knows only for the case at hand. The mathematician knows what she knows by pure thought, and since she has access to the axioms of Euclidean geometry, what she knows she knows for every conceivable case. (There is something in gender-neutral prose that seems to reflect the nature of reality.) But if the mathematician has access to every triangle, she does not have access to every theorem in the same way; no matter how many derivations have been done, there are always infinitely more to go. The truth simply has no end. This may suggest that Euclidean geometry is not only difficult but doomed. Not so. A process of sublimation is at work, one in which the mathematician does a part of the work involved in encompassing the infinite without ever doing it all. Although the axioms are finite, they are inexhaustible as well, allowing the mathematician to keep on going as long as she can keep on going.

The axioms of a mathematical system are among the artifacts of civilization, and it is to the unending work of cultivation and construction that Peano made his signal contribution. In 1889, he published a set of axioms for arithmetic, proposing for the first time since the ancient Greeks to encompass another aspect of the infinite within the purely human world of symbols, axioms, inferences, proofs, pencils, paper, and whatever else is needed to get the mathematician from one place to another.

There are five axioms:

1. 0 is a number
2. The successor of any number is a number
3. If *a* and *b* are numbers, and if their successors are equal, then *a* and *b* are equal.
4. 0 is not the successor of any number
5. If S is a set of numbers containing 0, and if the successor of any number *n* in S is contained in S as well, then S contains all the numbers.

■ ■ ■ ■ And here is Giuseppe Peano himself, moving slowly to the podium at history's great lecture hall. He is dressed in his habitually shabby brown suit. He speaks in his curious low raspy growl. He is addressing the mathematicians of the world, who sit in the audience, but like all great men, he is addressing the future. He feels the burden of explanation. There are furrows, deep lines of concentration, on his forehead. He stabs a short finger into the air.

"Zero," he says, "is a number."

"Just so," the mathematicians say. "Just so."

"Whatever the number," Peano rasps, "the next number is a number, too."

He pauses for a moment to see if everyone understands, then he adds: "The numbers form an eternal succession." Another pause. "They go on forever."

"Just so," the mathematicians say. "Just so."

"Whatever the numbers," Peano goes on, "if the same numbers come after them, the numbers are equal."

The mathematicians cough; they mutter and stamp their feet.

"No number before zero," Peano says bluntly. "The numbers may go on forever, but like the cosmos,"—and here Peano waves his hand toward the ceiling—"born soothsayers say in some explosion, they have a beginning."

"And, finally," Peano rasps, "properties of the numbers are inductive. They spread upward like a stain moving in steps. If zero has a certain property, and if the fact that any number has that property means that the number after that has the same property, then all the numbers have that property."

Finito.

■ ■ ■ ■

THE STAIRCASE OF ADDITION

The Peano axioms offer a resolution of infinitely many numbers into finitely many symbols. The resolution is generous enough to envelop all of the ordinary arithmetical operations. Addition is an example. If b is the number 1, $a + b$ is defined as the successor of a. The definition and the facts are in accord. The sum of 3 and 1 *is* the successor of 3. I might as well call the definition a rule—rule number 1, in fact—since the verbal contraption within which it is expressed is in essence imperative, serving to get someone to do something.

The operation that worked so well on 3 + 1 works well again on 3 + 2. We know, because the Peano axioms tell us so, that 2 is the unique successor of some other number—1, as it happens. Given the value of 3 + 1, the value of 3 + 2 may be computed by adding 1 to 3 + 1. The *sum* of 3 and 2 is the *successor* of 3 + 1.

The operation that has now worked so well twice, works for a third time with 3 and any number. The sum of 3 + *any number c* is the successor of 3 + *some number b*, where *c* is the successor of *b*.

With the anchor of the number 3 hauled up and discarded, the same operation that has worked so very well works with any two numbers. The sum of *any two numbers a* + *c* is the successor of *a* + *b*, where *c* is the successor of *b*.

And this is rule number 2.

Rules 1 and 2 are step-like in effect. The mathematician computes the sum of two numbers by first trundling down to the arithmetic basement, where he consults rule number 1, and then clambering back up again until he reaches the desired sum. Wishing to know the sum of 3 and 7, he follows the following staircase:

	3 + 7 = the successor of 3 + 6	3 + 7 = 10	
↓	3 + 6 = the successor of 3 + 5	3 + 6 = 9	↑
	3 + 5 = the successor of 3 + 4	3 + 5 = 8	
	3 + 4 = the successor of 3 + 3	3 + 4 = 7	
	3 + 3 = the successor of 3 + 2	3 + 3 = 6	
↓	3 + 2 = the successor of 3 + 1	3 + 2 = 5	↑
	3 + 1 = the successor of 3	3 + 1 = 4	

→ The successor of 3 = 4 →

Each step down and every step up is mediated by rules 1 and 2; and their engagement proceeds farther only by means of an exchange of identities.

Every magic trick depends on the displacement of attention. This one, too. Having counted to ten along their fingers, even students who are not prepared to count much higher may well wonder why a complicated definition is needed to determine something that has already been determined. As the magician's handkerchief flutters over his shabby top hat, the real rabbit may now be seen hopping toward the carrots. Let the rabbit go: here are the carrots. Something complicated—addition—has been defined in terms of something simple—succession. Something mental—addition—has been defined in terms of something mechanical—succession. Something infinite—addition—has been defined in terms of something finite—succession. Something derived—addition—has been defined in terms of something primitive—succession. No intuition is at work and were it not for the fact that those steps are being conducted in the context of the Peano axioms, which themselves retain moist traces of Peano's powerful intelligence, the mathematician might easily persuade himself that by an act of singular prestidigitation, he has somehow gained something for nothing.

A CHECKLIST

Leibniz turned to his Universal Characteristic and a checklist to ratify ordinary judgments about ordinary concepts. The Peano axioms and the definition of addition suggest the possibility of subordinating sums to the same checklist, the judgment that $3 + 2 = 5$ requiring thirteen steps before it emerges on the south—the sunny—end of certainty:

1. *Consider:* "3 + 2 = 5."	Check
2. *Look up:* "Peano axioms."	Check
3. *Consider:* "3 + 1."	Check
4. *Look up:* "rule 1."	Check
5. *Accept:* "3 + 1 = the successor of 3."	Check
6. *Accept:* "the successor of 3 = 4."	Check
7. *Accept:* "3 + 1 = 4."	Check
8. *Look up:* "rule 2."	Check
9. *Accept:* "3 + 2 = the successor of 3 + 1."	Check
10. *Look up:* "line 6."	Check
11. *Accept:* "3 + 2 = the successor of 4."	Check
12. *Accept:* "the successor of 4 = 5."	Check
13. *Print:* "3 + 2 = 5."	Check

The checklist embodies the staircase of addition, of course, but it goes farther, encompassing, in small, discrete steps, the *whole* of a process moving from *Consider:* "3 + 2 = 5" to *Print:* "3 + 2 = 5" and so wending its way from contemplation to conviction.

■ ■ ■ ■ Gottfried Leibniz now appears on the stage. A crude semaphore in hand, he is signaling frantically to Peano, whom he sees far in the future. Peano cups his ear and listens. "Leibniz," he says graciously, "stated two centuries ago the project of creating a universal writing in which all composite ideas would be expressed by means of conventional signs for simple ideas, according to fixed rules."

Signs, symbols, rules—some sixty years before logicians finally fixed the concept in formaldehyde, an aspect of the human mind had been reflected as an algorithm.

■ ■ ■ ■

THE DREAM MERCHANT

■ ■ ■ ■ It is somewhere, I am sure, that Juvenal relates the wonderful story of a quarter old in Rome when Rome was young, where an ancient Jewish family eked out a living selling dreams. Bystanders wandered in and out of the merchant's stall, passing the time, talking of dreams they might purchase. Workers and slaves stooped from labor asked timidly for dreams of wine and ease. Women asked for dreams of love, and men for dreams of women. Senators hustled themselves ostentatiously into the cramped quarter, their servants pushing aside the goats that still gathered in the streets, cropping the grass between the cobblestones, and demanded dreams of eloquence and power. Late at night, when Rome lay dark, the sleepless sidled along the narrow alleyways that led to the dream merchant's stall, their lanterns held aloft, and with their knuckles rapped on the dream merchant's shuttered window.

"Who's there?" he would hiss indignantly. "At this hour."

"I cannot sleep, I need a dream."

There was a pause as the merchant fumbled with his tunic, and then a clatter. The door opened. He stood there, the lantern throwing a flickering light on his narrow, ruined face.

"A dream you want? What kind of dream."

"Any dream. Any dream at all."

"For five denarii you can dream of milk and honey."

"I have two."

"Dream of goats and donkeys, then," said the merchant, closing the door slyly.

"And for three?"

"For three denarii? You can dream of Jerusalem."

"I am not a Jew."

"Beggars can't be choosers," observed the merchant.

"Very well. Let me dream of Jerusalem."

The clank of money filled the night air, and then the merchant closed his door.

One day in winter, when the Roman streets were filled with fog, and men hustled themselves from the baths with their tunics clasped around their throats, a small brown man dressed in the Greek fashion knocked on the door of the dream merchant's stall.

"I am here on behalf of my master," he said, speaking with the derisive accent of one whose native language is not Latin.

The dream merchant fingered his long snow-white beard.

"That you're here, I can see for myself," he said. "What is it that your master wants?"

"A dream."

"I have dreams and I have dreams."

"My master is wealthy."

"Let him dream of beauty, then," said the dream merchant. "For one hundred denarii he can dream that he occupies a palace made of beaten gold. There is the scent of incense in the air. He will lie on a bed of silk, underneath crushed violets, and women with dark black eyes will fan the perfumed air and sing for him."

The small brown servant withdrew a leather pouch from beneath his tunic and carefully counted one hundred denarii. The dream merchant accepted the money, held up his finger, and withdrew into the soiled interior of his stall. In a few moments he was back, carrying the dream of beauty.

The next day was the Sabbath. The dream merchant's stall was closed and shuttered, but the day after that, the small brown man was back. He knocked again at the door of the stall. The dream merchant regarded him with old hooded eyes.

"And so?"

"My master was very pleased," he said, "but now that he has dreamed of beauty, he wishes to dream of love."

The dream merchant nodded sagely and said, "For two hundred denarii, your master can dream that he has spent the night within the Temple of Love where the Goddess Aphrodite herself will bewitch him with her charms. He will couch to sighs of spring and taste the fruits of paradise."

Once again, the little brown man withdrew his purse from beneath his tunic and once again the dream merchant brought forth a dream.

A week passed in which the dream merchant sold dreams to soldiers, stiff from battle, and to women, who had given birth to stillborn children, and to fortune-tellers, oppressed by signs and symbols.

And then the little brown man was back. The dream merchant regarded him with his old hooded eyes.

"And so," he said again.

"My master was *very* pleased," he said. "But now that he has dreamed of beauty and of love, he wishes to dream of truth."

"Ah," said the dream merchant. "That is my most expensive dream. Few wish to dream of truth and fewer still can pay for it."

"How much is it, to dream of truth?"

The dream merchant paused, as if he were calculating the sum. Then he said brusquely, "If your master wishes to dream of truth, he must come here himself and I will give him the dream."

The servant withdrew.

The next day, there was a commotion in the quarter as a palanquin, preceded by four armed guards, elbowed its way through the district, causing goats and chickens to scatter in every direction. The palanquin stopped before the dream merchant's stall, and a tall, plump man of perhaps fifty wearing an immaculate tunic emerged and after motioning to one of his guards to rap on the dream merchant's door, stood there solemnly in the brilliant winter sunshine.

The dream merchant emerged, rubbing his eyes.

"I am Aristarchus," said the man, speaking in Greek. "I am here because I wish to dream of truth."

The dream merchant shrugged his shoulders and then rubbed the forefinger and thumb of his right hand together.

Aristarchus raised his eyebrows in order to ask the price.

"One thousand denarii," said the dream merchant.

Aristarchus appeared to hesitate, whereupon the dream merchant began to close the door to his stall.

"Now understand me," Aristarchus said quickly. "It is not the thousand denarii." He gestured to his expensive palanquin and to the bodyguards standing patiently beside it.

"What then?"

"I have read the philosophers," said Aristarchus slowly, "and I have listened to the soothsayers and I have spoken to the priests who know the Eleusinian mysteries, but I have never seen the truth. Will I see the truth if I dream this dream?"

The dream merchant shrugged his thin shoulders underneath his stained caftan. "Even in dreams, no man sees all of the truth."

"What part will I see?" Aristarchus asked.

"The part that you *can* see," replied the dream merchant.

Aristarchus stood for a moment lost in thought, and then, making up his mind, motioned to his servant, who had been standing quietly beside the palanquin, to fetch the thousand denarii.

The dream merchant accepted the coins gravely and withdrew into his stall. Fifteen minutes passed when at last the dream merchant emerged with the dream of truth, which he placed in the hands of the servant. Aristarchus nodded and stepped back to his palanquin.

A day passed and then a week. On the day following celebrations at the Temple of Jupiter, there was again a commotion in the quarter where the dream merchant had his stall and again Aristarchus's palanquin, surrounded by his personal guards, made its way through the narrow streets.

The palanquin stopped with a clatter before the stall. Aristarchus emerged, preceded by his servant. He motioned to his servant to fetch the dream merchant.

After a few moments, the dream merchant appeared. He looked at Aristarchus standing in the sunlight and said, "Good morning," quietly.

Aristarchus said, "I dreamed the dream of truth for seven nights."

"And?"

"Each night I dreamed that I was mounting a series of broad white steps, like those at the great Temple of Jupiter."

Aristarchus paused, as if he were collecting his thoughts. Then he continued, "At first, my heart beat violently in my chest. I could see that with each step I was getting closer and closer to the truth. I was filled with a great longing to see the sun."

The dream merchant looked quizzically at Aristarchus.

"I climbed higher and higher, until my legs began to ache."

From the small narrow streets around the dream merchant's stall came the squawking sounds of a Roman morning. Women were shouting to one another and the cries of children and chickens filled the air.

Aristarchus looked at the dream merchant. "As I climbed," he said, "I could feel the warmth of the sun emerging. Brightness fell from the air. The very steps glistened beneath my feet. A great feeling of happiness spread through my limbs."

The dream merchant continued to look at Aristarchus with his old hooded eyes but said nothing.

"It was then that I awoke," Aristarchus said. "The day was gray and I felt as if I were stepping into a cold bath."

"The next night, I dreamed the dream of truth again, and again I found myself mounting the same series of steps. This time I rose higher than before and again I saw brightness fall from the air."

A small, sly smile played across the dream merchant's face.

"And again I awoke," said Aristarchus, "and again the dawn was gray. For seven nights, I dreamed the dream of truth, and for seven nights, I climbed and climbed until I awoke, and for seven nights the dawn was gray."

"And?" asked the dream merchant.

"I am no closer to the truth than I was," said Aristarchus. "I sense its radiance, but I cannot reach it."

"You must dream the dream again," said the dream merchant.

"My dreams are valuable," said Aristarchus peevishly. "If I dream the dream again, when will I reach the truth?"

"When you have climbed all the stairs," said the dream merchant.

"And when will I have climbed all the stairs?"

"When you have reached the truth."

Aristarchus stood irresolutely in the spreading sunlight. Finally, he said, "That is not a very satisfactory answer."

"Yours was not a very satisfactory question," replied the dream merchant.

"It was not the dream I thought to dream," said Aristarchus.

The dream merchant spread his hands wide. "Nonetheless," he said, "it was the dream you dreamed."

For a long moment, Aristarchus stood quietly in the sunshine, as if he were thinking of what to say; finally, he motioned to his servant, who had been standing quietly by the palanquin, and in Greek told him to fetch the dream.

"At once, Master," said the servant, disappearing into the palanquin and reemerging with the dream.

"I am returning your dream," Aristarchus said gravely. "It is a dream I no longer wish to dream."

The dream merchant nodded his head as if to say that he understood that Aristarchus would have his money back but was too proud to ask. "It is expensive to dream of truth," he said. "By comparison, beauty and love are cheap."

For the first time, Aristarchus smiled broadly, revealing his even white teeth. He said something in rapid-fire Greek to his servant, who then presented the dream of truth to the dream merchant. "After all," he said, "money is only money."

"And truth is only truth," said the dream merchant.

"Yes," said Aristarchus.

And with that, he turned and reentered his palanquin, stooping low so that his head did not hit the carriage roof. There was a clatter and the palanquin proceeded down the narrow lane in front of the dream merchant's stall, preceded by its guards, who shooed children and chickens and the occasional goat from the way.

The dream merchant watched the procession recede until it had disappeared from sight. Just then, a young man wearing an immaculate tunic emerged from the lane that led to the dream merchant's stall. He had bright shiny eyes that were framed by the thick oiled hair that lay closely about his dark face. It was the poet Catullus.

He nodded courteously to the dream merchant, whom he knew very well, and said: "My Lesbia, I wish to dream of her again."

"Young men," he said, "wish always to dream of what they have lost."

Catullus looked at him curiously. "And old men?"

"Of what they have not found," said the dream merchant, turning on his heel to fetch the poet's dream.

■ ■ ■ ■

APPOINTMENT IN TURIN

Peano was a gifted teacher, but his career was marked by student riots of the sort that might have occurred at Berkeley or, later, at the Sorbonne. The accounts are delicious, the riots precipitated, as these things often are, by some trivial inconvenience. During the year of his official promotion to a professorship in mathematics, students took to the streets at the beginning and the end of the year, surging through the academic buildings, breaking doors, and sacking files. As far as I can determine, they were never criticized or punished, officials at the university and within the Turin city government taking the good-natured but obviously lunatic view that students are best trained by indifference.

He enjoyed returning to the village in which he was born. The farm is still held in the family name. His lean features and narrow feral face suggest the immemorial aspect of the peasant, an aspect that can now be found only in the sunbaked villages of the Abruzzi. He was a sophisticated mathematician and a member of the broad and generous European culture that to those living within its embrace seemed genuinely to have been marked by immortality. It soon lay in ruins, the common culture that it expressed hopelessly smashed. Peano lived until 1932, growing old under Mussolini's regime. Leibniz never having lost his hold on his imagination, he absorbed himself in various schemes for a universal language. He was productive almost until the end, but many men and women outlive their time, and I suppose that Giuseppe Peano was one of them. His chronic cough grew worse, his voice hoarsening into a croak. Still, he lived on and on, eating the same meals each day and each day trudging the steps to his apartment, the hallway smelling of garlic and smoke.

I followed his trail just once. I had driven all day and then all night, aimlessly wandering through southern France and then central Italy, crossing the border somewhere south of the Alps and then cruising the long smooth *autostrada,* a tape of Beethoven's Ninth Symphony playing on the tape deck of the little BMW convertible I was driving. Heavy trucks on the road. The smell of petrol and asphalt and from far away something tangy, rice fields under cultivation. Then the *autostrada* emptied itself into a network of suburban highways that by and by gave way to city streets. I was in Turin. Rain falling lightly, the city tan and brown. Long boulevards, dilapidated cafés and bistros, where men sat huddled over card tables, drinking Pernod with water, here and there a grocery store and even a few restaurants, the boulevard rounding a traffic circle at whose center stood a statue of Garibaldi. I spent the night at the Hotel Casanova, quite an unromantic place, the room low-ceilinged, the place smelling strangely of sawdust, some residue of the filler in the walls, I imagine.

It is Peano's city, Turin, but it was also Primo Levi's city, the place where he matriculated in chemistry and from where he was sent to Auschwitz. His own apartment building is not all that far from the building in which Peano lived and died. I knew the street and wandered there in the light rain; I pushed open the heavy wooden doors, triggering the hallway lights. I could see the central spiral staircase from which Levi had thrown himself to his death. After a short time, the hallway lights turned themselves off.

The Greeks were right about this, as they were right about everything. No escape.

CHAPTER 3

Bruno the Fastidious

Something in the universe establishes—it *guarantees*—that problems banished in one place sooner or later pop up in another, grinning and ineradicable. Tiring of domestic discontent, a man discovers upon divorce that loneliness is insupportable; the pain that has lately vacated the big toe takes up residence in the lumbar region, where somehow it seems worse. I am giving examples of a law of nature implacable as gravitation. The Peano axioms provide a route to the infinite. But if the axioms are finite, and so, too, each step on the arithmetical staircase, the *number* of theorems, steps, and conclusions that can be derived from the axioms is itself infinite again, if only because there are infinitely many numbers, and so inevitably questions about an arithmetical system reappear in disguise as questions about an axiomatic system.

Is the axiomatic system *consistent*? Can the mathematician be certain that sometime in the thirty-third century, some ingenious student in Calcutta, ascending the arithmetical staircase, might not discover that while "3 + 2 = 5" follows from the Peano axioms, so does "3 + 2 = 6," thus reducing the enterprise to incoherence? Can he or she be certain, for that matter, that the arithmetical system is *complete*? The staircase of arithmetic goes on and on: quite true; but as a young woman from Shanghai is destined to write (in Chinlish, as it happens), *It is big-odd that there are certain arithmetical too-truths that the staircase never reaches—the fact that* 3 + 17,293,456 = 3 + 17,293,459, *to be much-precise*. There follows the Chinlish sign for *huh?*

Sign, symbol, and staircase provide sanctuary against arithmetical doubt; and the examples that I have fabricated are absurd, if only because we know, and know with complete confidence—right?—that the sum of 3 and 2 is not 6, and that 3 added to 17,293,456 is just what it is supposed to be, the Peano axioms, and common sense, yielding 17,293,459.

But if the examples are absurd, not so the insecurity they engender. *That* remains, like the taste of tea. These examples are absurd. Very well. What of others? Who knows what Shambling Other might well be shambling up the arithmetical staircase? Or shambling down?

The eye of doubt has changed its focus, withdrawing its attention from the arithmetical system, and with the same old evil intensity has begun looking over the axiomatic system instead.

THE GNOME OF LOGIC

Aristotelian logic is massive and marmoreal, but every monument accumulates graffiti, and the syllogism is no exception. Medieval scholars discovered subtleties in the system and wrote their names in chalk to tell the world of what they found. Abelard was a logician as well as a lover, and while the riffraff in the twelfth-century Paris taverns were singing his love songs, banging their tankards on the table at every chorus, he was engaged in disputations—when not otherwise engaged in kissing Héloïse—lecturing on the Parisian hilltops, where students gathered at his feet.

Leibniz enlarged the margins of the Aristotelian system, and two centuries later, so did the English logicians George Boole, Augustus De Morgan, and John Venn, Venn especially providing a very nifty series of diagrams showing that if the circle of dogs is contained within the circle of mammals and the circle of mammals is contained within the circle of animals—I am sketching this on the blackboard now, and whoever is snoring in the back row better stop—*but just look at this,* the little circle is in the big circle and the big circle is in the bigger circle and all dogs *are* animals...

Never mind. It is *modern* logic that is the real stuff, and the real stuff is almost entirely the creation of Gottlob Frege, the gnome of logic.

His life was bleak. Frege was born in 1848 in Wismar, which lies in the province of Mecklenberg-Schwerin. Bleak enough. This is northern Germany, the land facing the Baltic Sea. Bleaker still. It is a countryside of dark and gloomy forests, hags and elves and goblins and toadish-looking men behind the somber trees. At night, the horned owls hoot and black-footed wolves trot restlessly along the forest paths and hunchbacks gather in darkened glens to play the clarinet.

Frege spent his entire academic career at the University of Jena, trudging like Peano in Italy up the obligatory steps of the academic ladder: a *Privatdozent* in 1871, and so authorized to accept students without pay, an *ausserordentlicher Professor* in 1879, a *Professor* in 1896, and thereafter a *Herr*

Professor, the open-voweled *Herr* followed in conversation by the three even beats of *Professor*.

He was married for many years—happily, so far as I know—*die gnädige Frau Frege* dying along with Europe during the course of the First World War, and so darkening a personality that was already dark, lonely, crabbed, solitary, and withdrawn.

And he seems to have been—in plain fact, he was—a ferocious anti-Semite, seeing in Germany's sad, doomed, cultured German Jews an alien and unwanted presence, and, no doubt, regarding the turbulent wave of eastern European Jewry, which had washed over Germany early in the century and with outstandingly bad judgment come to rest in Leipzig or Dresden or in Weimar itself, with feelings akin to frank revulsion. Disliking Jews, Frege disliked Catholics as well, the ink of his indignation ecumenical in its nature. He was deeply devoted to the German monarchy, its preposterous and dangerous kaiser receiving from Frege the respectful sentiments that he had nowhere else to discharge. With the exceptions of Bertrand Russell, Ludwig Wittgenstein, and, to a certain extent, Edmund Husserl—not an inconsiderable trio, of course—contemporaries could not fathom his work. It was ignored when it appeared, and if philosophers and logicians now agree that Frege was the greatest of mathematical logicians, if only because he was the first, their encomiums came too late to afford him solace. He died in 1925.

And he died alone.

A CALCULUS OF SHAPES

An axiomatic system establishes a reverberating relationship between what a mathematician assumes (the axioms) and what he or she can derive (the theorems). In the best of circumstances, the relationship is clear enough so that the mathematician can submit his or her reasoning to an informal checklist, passing from step to step with the easy confidence that the steps are small enough so that he cannot be embarrassed nor she tripped up. It is in the context of that checklist that the concept of an algorithm moves crabwise onto the screen of consciousness, but still gray, still only half-formed, still fuzzy and indistinct. That checklist, after all, is my own purely rhetorical contrivance, and the logician prepared to scruple at some arithmetical or logical claim is hardly apt to be reassured by my own booming "check."

"Whad'ya mean check? Go check yourself."

Her objection is, of course, that *I* have left too much unspecified, the whipcrack of my checks timed suspiciously to sound when I need them to sound.

It is within the context of a *formal* system that the checklist is itself absorbed into the structure of the system, like paint seeping into walls, the specification of a formal system providing the first clear example of an algorithm, and so bringing an old concept to life.

An axiomatic system comprises axioms and theorems and requires a certain amount of hand-eye coordination before it works. A formal system comprises an *explicit* list of symbols, an *explicit* set of rules governing their cohabitation, an *explicit* list of axioms, and, above all, an *explicit* list of rules *explicitly* governing the steps that the mathematician may take in going from assumptions to conclusions. No appeal to meaning nor to intuition. Symbols lose their referential powers; inferences become mechanical.

The propositional calculus is the simplest imaginable formal system, and as its name suggests, it is a system in which whole propositions (or sentences) come to traffic with one another. The system is spare enough so that only a handful of symbols suffices.

A set or collection of propositional symbols:

$$P, Q, R, S, \ldots$$

to begin with. These represent sentences, with *P* standing impassively for "The liver is a large organ," *or* "Warsaw is in Poland," *or* any other proposition. No decomposition of propositions into their parts is permitted. "The liver is a large organ" and "Warsaw is in Poland" carry on like stiffly moving mummies, the logician indifferent to the substance beneath those shrouds. With internal configurations wiped away or ignored, all that remains to distinguish propositions is their truth or falsity.

A collection of special symbols next:

$$\sim, \&, \vee, \supset$$

They are "$\sim$" (the symbol for "not"), "&" (the symbol for "and"), "$\vee$" (the symbol for "or"), and "$\supset$" (the symbol for implication, read, "if... then"); their purpose is to stand like sentinels between or in front of propositional symbols.

Finally, two marks of punctuation:

(,),

a left parenthesis, and a right parenthesis.

The outer shell of the proposition that the liver is a large organ *or* that Warsaw is in Poland emerges on this side of sanity as $(P \vee Q)$.

Nothing in this is yet especially formal, nor, for that matter, especially interesting; symbols have been created to represent the shells of sentences. But with the symbols in place, the logician now undertakes what will become a characteristic maneuver, subtracting from the system all traces of its meaning or interpretation. The remainder *is* a formal system, a scheme of symbols whose identity goes no farther than their shapes.

That bright, prophetic voice now sounds again, churning through the centuries. "Combinatorics," Leibniz wrote (in remarks I have already cited), "treats of calculus in general, or of general signs or characters (such as *A, B, C,* where any one could be taken for another at will), and of the various laws of arrangements and transitions, or of formulas in general."

RIGHT BRUNO? RIGHT?

The specification of a formal system proceeds as if the logician were explaining things to an able but malevolently punctilious interlocutor—someone named Bruno, say.

The list of primitive symbols having already been presented to Bruno, who has, I am assuming, accepted them with a reluctant grunt, the logician next attends to the grammatical rules governing their formation. There are only three. (Bold capital Roman letters such as "**A**" or "**B**" are used to talk *about* formulas and represent an explanatory device. They are not themselves a part of the formal system.)

1. A propositional symbol standing alone is grammatical or well-formed.
2. If any formula **A** (such as $(P \vee Q)$) is well-formed, then so is its negation ~**A**, (~$(P \vee Q)$ in this case).
3. If **A** and **B** are well-formed formulas, then so are (**A** & **B**), (**A** ∨ **B**), and (**A** ⊃ **B**).

These rules remind Bruno, and should remind *you,* of the arithmetical staircase in which action proceeds by means of the double trigger of descent and verification.

Bruno interrupts to ask for a demonstration. I am happy to oblige.

1. $\sim((P \vee Q) \,\&\, R)$	←	(Is this formula grammatical?)
2. $((P \vee Q) \,\&\, R)$	←	(Yes, if *this* formula is grammatical, by rule 2.)
3. $((P \vee Q) \,\&\, R)$	←	(Is this formula grammatical?)
4. $(P \vee Q)$	←	(Yes, if this formula is grammatical, and
5. R	←	this formula is grammatical as well, by rule 3.)
6. $(P \vee Q)$	←	(Is this formula grammatical?)
7. P	←	(Yes, if this formula is grammatical, and
8. Q	←	this formula is grammatical as well, by rule 3.)
9. P	←	(This formula *is* grammatical, by rule 1.)
10. Q	←	(Ditto.)
11. R	←	(Ditto.)

→ and so up the staircase to verify line 1 →

I have interpolated my own parenthetical remarks strictly for Bruno's convenience. The question whether a given formula is well-formed is decidable in a finite number of steps and decidable by mechanical means and decidable by means of a procedure that defies even Bruno's capacity to scruple and to doubt.

Right Bruno? Right? Right.

PROOF BEYOND DOUBT

A formal system has been given hands, eyes, and ears; no unwanted webbing between the toes—*check*. No extra digits anywhere—*check*. But before the shiny new creature can get from axioms to theorems, it must first have a set of axioms to work with and explicit rules for their manipulation. Otherwise the system would not be formal.

There are three axioms.

1. $P \supset (Q \supset P)$
2. $S \supset (P \supset Q) \supset ((S \supset P) \supset (S \supset Q))$
3. $(\sim P \supset \sim Q) \supset (Q \supset P)$

And there is no point—*absolutely none*—in asking what these symbol strings *mean.*

They don't *mean* a blessed thing. They function as shapes and so as symbols reduced to their essence.

The rules of inference next. There are only two, but the second requires an explanatory detour. The sequence of shapes

$$(P \supset Q)$$

consists of five symbols; the sequence of shapes

$$(R \supset S) \supset (T \supset W),$$

of eleven. But plainly, these two formulas have the same form, the horseshoe of implication ("⊃") standing in both cases between two distinct formulas. The second formula, logicians say, is obtained from the first by *substitution,* with "$R \supset S$" substituted for "P" and "$T \supset W$" for "Q." Substitution proceeds apace only when like formulas are substituted for like propositional symbols and different formulas for different symbols.

Substitution in hand, the logician is now prepared to express the rules of inference.

1. If the formulas **A** ⊃ **B** and **A** are either axioms or theorems, then the formula **B** is a theorem, too.
2. If the formula **A** is an axiom or theorem, and if substitution takes **A** to **B**, then the formula **B** is a theorem as well.

Bruno having been appeased, the formal system is now finished and ready to be put to use, say in proving that $P \supset P$.

And I do mean *proof,* or entry into the absolute, investing the word not only with its ordinary meaning, but with the whole weight of conviction—conviction beyond doubt, conviction beyond the possibility of doubt, and conviction even beyond the possibility of even the possibility of doubt.

The proof proceeds by steps in which every line is either an axiom or follows from an axiom or follows from what follows from an axiom. Having proceeded according to the rules of inference, an argument constitutes a proof of its very last line.

1. $S \supset (P \supset Q) \supset ((S \supset P) \supset (S \supset Q))$	(axiom 2)
2. $P \supset (Q \supset P) \supset ((P \supset Q) \supset (P \supset P))$	(substitution of P for S, Q for P, and P for Q)
3. $P \supset (Q \supset P)$	(axiom 1)
4. $((P \supset Q) \supset (P \supset P))$	(derived from lines 2 and 3 by rule 1)
5. $P \supset (Q \supset P) \supset (P \supset P)$	(substitution of $(Q \supset P)$ for Q)
6. $P \supset (Q \supset P)$	(axiom 1)
7. $P \supset P$	(derived from lines 5 and 6 by rule 1)

That is all. The formula $P \supset P$ is demonstrable. Every proposition implies itself, and while this is something that everyone antecedently believes, now it is something known. I have just proved it.

And Bruno? Bruno is content.

DOUBLE VISION

Let me tell you. We logicians get up in the morning, groan with dissatisfaction, brush our teeth and comb our hair, and then waddle off to face the newspaper, coffee, and those miserable bran muffins that in a halfhearted endeavor to lower our cholesterol, we propose manfully to choke down each and every morning. We live in a world of meaning, purpose, intention, and commitment, where every symbol that we use, we use with some interpretation in mind, "I love you, Daphne," meaning that *I* love *you,* Daphne, and so comprising more than a sequence of fifteen shapes. *Much* more.

The intended interpretation of the propositional calculus is no mystery. A single propositional symbol is either true or false, the panoply of possibilities displayed by a truth table:

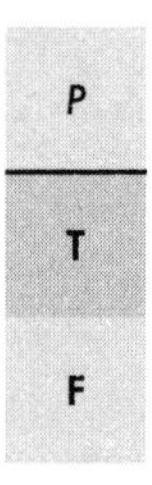

P
T
F

For purposes of display, one symbol requires two lines; two symbols, four; three, eight; four, sixteen; and n symbols, 2^n lines.

Containing only two propositional symbols, the wedge of disjunction thus takes a truth table of four lines:

P	Q	P ∨ Q
T	T	T
T	F	T
F	F	F
F	T	T

each expressed as a hypothetical: *if P* is true (or false) and *Q* is true (or false), then $P \vee Q$ is true (or false).

The negation of a proposition *P* is true only if *P* is false, and vice versa. No surprise, that. No surprise either when it comes to conjunction: *P* & *Q* is true if *P* is true and *Q* is true.

The otherwise smooth fabric of this analysis encounters a snag at the horseshoe of implication:

P	Q	$P \supset Q$
T	T	T
T	F	F
F	F	T
F	T	T

The first and second lines are straightforward, but the third and fourth engender a suspicious scruple. Just why should the logician reckon the proposition "*If* Warsaw is in China *then* the liver is a large organ" true simply on the grounds that Warsaw is *not* in China? Why acquiesce in the hypothetical that *if* Warsaw is in China *then* the liver is a *small* organ?

From time to time, logicians try to explain themselves on this score, but to no avail. The truth table for this connective is simply *arbitrary,* the logician having made a decision rather than a discovery.

Nothing more is at issue. Trust me.

ABSOLUTE TRUTH ABSOLUTELY

Truth tables and the connections they reveal belong to the world beyond the world of symbols; and apart from their appeal as small, compact, urbane, and diamond-bright devices, they are useful in creating a new concept. For the most part, the formulas of the propositional calculus are true under some distribution of truth to their constituents, and false under others—witness $P \vee Q$; but the *tautologies* are true whatever the distribution and so absolutely true. The formula $P \supset P$, lately the conclusion of a proof, is a tautology as well, a circumstance revealed by yet another truth table:

P	P	$P \supset P$
T	T	T
T	F	Impossible
F	F	T
F	T	Impossible

There is simply no way to assign truth-values to P that suffices to falsify $P \supset P$; if the absolute seems somewhat less prepossessing than one might have wished, this is only because like all deities, this one has chosen to reveal itself by means of trivialities.

The tautologies lie beyond the purview of a formal system, the proofs within. The logician has access to both concepts and so enjoys a position from which he can look on two worlds. Still, if the idea of *unassailable* proof resides within a formal system, something prooflike lies beyond, and that is the ordinary human practice of showing that something or other must be true, because look, if this is true, then this must be true, and this is true and so that is true as well.

Theorem and proof figure again in this *in*formal discourse, but they figure in their ordinary sense, a proof coming to mean roughly what the logician can get away with, and a theorem what she has gotten away with. This is hardly cause for criticism. The logician does what she can.

And what she can do is draw the curtain of a connection between two concepts, establishing in her own vernacular—and ours—that every theorem is a tautology and every tautology is a theorem. Note the double bridge: every *theorem* is a *tautology* and every *tautology* is a *theorem*. A part of the logician's argument is simple and striking, although hardly proof against any conceivable form of doubt (down, Bruno, down). The axioms of the propositional calculus are all tautologies. *Check*. The rules of inference take tautologies to tautologies and nothing else. No check yet, but after some reflection, the logician's *check* is forthcoming. Call it a promissory check, one promising to prove on demand that every theorem is a tautology.

It is somewhat more difficult to construct an argument showing that every tautology is a theorem, but it can be done and I can do it. My check is in the mail.

With those particular checks in hand or en route, the logician has everything needed to finish up the argument, and the propositional calculus is complete.

The consistency of the propositional calculus follows from its completeness. If the calculus were *in*consistent, anything whatsoever could be made the subject of a proof. An inconsistent system is intellectually profligate. But the formula $P \supset {\sim}P$ is not a tautology, and so it is not a theorem. Only two checks are required to check this out.

And more. The propositional calculus is *decidable* as well as complete and consistent. In the world *beyond* the formal system, there is a finite, explicit, and effective scheme for determining whether an arbitrary formula is a theorem *within* the formal system. Is the formula a tautology? If so, it is provable, and if not, not. The logician need not derive $P \supset (P \vee Q)$ from the axioms in order to prove that $P \supset (P \vee Q)$. A truth table suffices, and none of that business of descent and verification. Unlike the Peano axioms, which for anything I have said are neither complete, consistent, nor decidable, the propositional calculus has all the modern conveniences.

THE EPONYMOUS BRUNO

Bruno has played a purely rhetorical role in this discussion, at once a figure of fun and a focus of doubt; but not long after I amused myself with his creation, I realized after a certain reorganization of my own memories, that Bruno had a living model, a graduate student at Princeton, in fact, named Daniel Mesmeister. Looking somewhat like a cross between Ichabod Crane and Basil Rathbone, Mesmeister surveyed the graduate-school commons and the dining room itself with dark, penetrating, altogether remarkable eyes. He had no special fund of knowledge: no very remarkable argumentative gifts, nor any particular source of personal magnetism. His great talent lay in his ability to convey some primitive sense of skepticism by means of an attitude which resolved itself into a stare.

He rarely had anything very interesting to say and what he did say he generally said with a view toward dismissing both sides of every argument, thus criticizing American intervention in Vietnam and with a snort of derision objecting to American withdrawal as well. There followed that stare, blank and intimidating. I suppose that every graduate school has its own Mesmeister, but ours was singular in his calculated decision defiantly to occupy the whole of a skeptical space that the rest of us were content to explore only from time to time.

One day, Mesmeister disappeared from the graduate school. I had no idea where he went or what had become of him. Having resented his presence, no one regretted his absence. Years later, an old friend from graduate school sent me the following short story, with the recommendation that I make of it what I would.

The window had for some reason not been opened; the room was humid and the air had the sweet, damp, unattractive smell peculiar to the late hours of the night. Leo Rubble, his legs splayed to form a triangle, the sole of one foot resting on the knee of the other, slept on his back, his hands cupped underneath his belly, a slight childish smile on his lips. A young woman slept on the quarter of the bed that he had not made unusable. She was curled into the fetal position, her palms pressed together, her hands between her thighs. Her very fine blonde hair had fallen sideways over her face, obscuring her features. She was not snoring precisely, but breathing with a kind of wetness, the sound beginning with a ragged inspiration and ending with a muted but explosive puff.

The light on the white Princess telephone beside the bed glowed for a moment; the telephone rang.

Leo snorted and sat up partway, his torso supported by his elbows.

"Telephone," he said thickly.

The young woman sat upright in bed with a great feathery movement. "It's Richard," she said. "You get it. I'm not here."

The telephone had rung twice. Leo picked up the receiver and rotated his torso so that he was lying entirely on his side. He cleared his throat.

The young woman reached over to place her left hand over his forearm; she mouthed the word "who."

"Danny!" Leo said; directly afterward, he sat up again, the telephone in his lap.

The young woman flopped backward on the bed, her eyes closed, her hands on her cheeks. She pressed her lips together and bit them until the blood had drained, leaving them white.

"Look at me," she said, taking her hands from her cheeks and holding them out. "I'm shaking like a leaf."

Leo shook his head vigorously from side to side to indicate that he was being distracted.

The young woman turned on the bed to look at the clock mounted in the radio. "Does he have to call in the middle of the night?"

Leo covered the telephone receiver with his hand, and whispered, "Sorry," to the young woman; then he made a writing motion to indicate that he needed paper and a pen. The young woman, who had continued to lie on her back, her eyes open and staring, rolled onto her side and reached for the spiral-bound notebook that lay on the nightstand beside her side of the bed. A ballpoint pen had been inserted in the spirals.

Leo opened the notebook.

"What?" said the young woman.

Leo shook his head.

The young woman sat up; she reached for her pillow and placed it in her lap. She began to fluff the pillow.

Leo covered the receiver again, and whispered, "Wait."

"I *am* listening," he said to the telephone.

"I *am* listening," said the young woman, in a mean mimetic whine.

"Danny," said Leo. The sound of his voice was heavy. For a time, he did not speak. Then he said "Danny" again and then "yes."

He replaced the receiver on its cradle and folded his hands together.

The young woman, who had remained sitting, the fluffed pillow in her lap, looked at him with her eyes narrowed slightly.

"What?" she said.

Leo Rubble sat with his hands still folded, the telephone on the bed beside him.

"Well?" said the young woman.

"He's calling everyone," said Leo.

"Why?"

"To let them know."

"To let them know *what*?"

"That they're planning to steal his eyes. There are these people that are planning to steal his eyes."

For a moment the young woman said nothing.

"Oh dear," she finally said.

■ ■ ■ ■

CHAPTER 4

Cargoload and Crack-Up

Imagine a straight line going forward into the future, conveying a cargoload of accelerating ideas from roughly 1890 to 1931, the algorithm among them. That cargoload is due to crack up shortly, but not before the algorithm has hopped off, safe, sound, but inevitably somewhat bruised. Cargoload and crack-up encompass a great, tragic human story, and like all great stories, this one contains elements of arrogance and reprobation, crime and punishment, some entirely human expectation that in the establishment of mathematical certainty, the human heart would discover not only relief but an encounter with the absolute. But for the moment, that lies in the future. It is yet 1900 or so. No one knows what is to come.

THE MASTER OF INFERENCE

The propositional calculus is an example of a formal system and therefore an example of an algorithm. But an example is not an explanation, still less a definition, and in any case the propositional calculus is an instrument of vibrant triviality. It was arithmetic itself that Frege meant to master, and for this, a calculus of propositional shapes is entirely beside the point. What is the use, after all, of a system in which "$2 + 2 = 4$" and "$\sqrt{36} = 6$" are simply swallowed up as *P* and *Q*? I might as well answer my own question. No use at all.

Nor is the Aristotelian syllogism of much use, however much it may have entered the late nineteenth century embroidered by diagrams. The Peano axioms talk about numbers, some numbers, all numbers, and the properties of numbers; the transmogrification of Peano's mathematical vernacular into a formal system requires a system of notation and a scheme of inference that simply does not exist within the ambit of Aristotelian logic. This system of notation Frege created is what has come to be called the predicate calculus. And it too survived the crack-up to come. It is now his monument.

The predicate calculus begins with a token nod toward old-fashioned grammatical analysis in which simple sentences such as "John is praying" are decomposed into subject—"John"—and predicate—"is praying." Token though the nod may be, it nonetheless represents a power in prospect that goes entirely beyond the propositional calculus. The propositional calculus deals only in propositions; the predicate calculus slices propositions into their constituents and thereafter makes for entirely new ground.

Within elementary algebra, that most moanful of high-school subjects, variables such as x, y, z, . . . are used to designate numbers so that Mrs. Crabtree, now and forever the embodiment of high-school teachers everywhere, can affirm not only that $5 \times 5 = 25$, but that $x^2 = 25$, where x^2 has something of the force of an ordinary English pronoun—*it* times *it* is 25. The specificity of the number 5 is lost in the expression $x^2 = 25$, but recaptured on manipulation, the number's identity appearing from an alembic of algebraic constraints. *It* is *that* number *which* . . . And thereafter Mrs. Crabtree smooths the ruffled fold of her white blouse and sighs.

For more than three hundred years, elementary algebra was given over to elementary numerical manipulation; but there is no reason—is there?—that variables must be tied to numbers. The individual variables x, y, z, . . . now make a semiformal appearance, performing the function in logic that pronouns perform in ordinary English, the sentence "She is blonde," cognate to the proposition "x is blonde," both *she* and x specifying something but specifying that thing indeterminately.

Individual variables stand for men or mammals, asteroids or astronauts, politicians or prigs, the elements, in fact, of any universe of discourse, the very phrase, "a universe of discourse," signifying a new direction in thought, one in which the old-fashioned out-there universe of astronomers and astrologers is replaced by a new-fashioned universe of signs and symbols and the things that signs and symbols signify or mean.

Individual variables constitute one-third of a conceptual scheme; predicate symbols, the second third. These are signified by uppercase Roman letters, F, G, H, . . . , and correspond to ordinary English predicates— . . . is *blonde*, *bold*, *beautiful*, *doomed*. The proposition that x is blonde now vanishes into symbols along with that blonde: Bx.

Now being blonde is something that one person is or flaunts: the predicate has one place reserved for one individual; but *loving* or *leaving* are relationships between two individuals, the requisite predicates require two objects: Irma *loves* Philip, whereupon x loves y, whence Loves(x, y), whereupon Irma *leaves* Philip, whence x leaves y, whence Leaves(x, y). The predicate calculus encompasses relations as well as predicates, an additional range of symbols standing for two-place relationships (loving, leaving), three-place relationships (being between so and so and such and such, as when Robert is *between* Philip *and* Irma), and so on up toward *n*-place relationships, where any number of individuals are coordinated by a single multiheaded relationship, as in an aboriginal clan. With individual variables and the full range of predicate symbols in place, the predicate calculus is capable of representing the dark interior of a great many previously inaccessible sentences. That horse's head that stopped the syllogism in its tracks? Nothing more than y is a horse & x is the head of y.

The beat of creation has rumbled twice, and, *yes,* the construction of the predicate calculus is an act of creation, the logician managing, by means of various dry acts of specification and stipulation, to endow bits and pieces of his own mind—symbols, after all—with all the lunatic energy of life itself. That mysterious beat need rumble one more time. Quantifiers now come into play. In ordinary English, quantification is expressed by "some" and "all." *All* men live in fear, but *some* in dread. These dark adverbs are symbolized in the predicate calculus by the universal quantifier, $\forall$, and the existential quantifier, $\exists$. They function by operating on variables (hence their technical description as variable-binding operators). Someone ate an albatross? Very well. There is an x such that x ate an albatross. In symbols: $\exists x\mathrm{A}x$. Did everyone eat that albatross? For every x, x ate an albatross. In symbols again: $\forall x\mathrm{A}x$. In these constructions, the quantifier *binds* the variable, exerting its pull and its powers with respect to the variable which it flanks, and this over the whole of the formula just beyond the quantifier. The variable x is bound in $\forall x\mathrm{A}x$; it is bound as well in $\forall x(\mathrm{A}x \mathbin{\&} \mathrm{G}y)$, but the variable y floats forlorn and free in the same formula, beyond the scope of the universal quantifier, which is busy managing x. Where necessary, parentheses mark the limits of the quantifier's binding powers.

The apparatus of quantification makes for an interpretation of generality that goes beyond the Aristotelian syllogism. The syllogism places the proposition that all dogs are animals in something like a generic shell: all *A*s are *B*. The predicate calculus cracks open the shell to reveal its hidden hypothetical: *If* anything is a dog, *then* that thing is an animal. Using quantifiers, variables, and a propositional connective, the hypothetical emerges pale and pink: $\forall x(Dx \supset Ax)$.

The universal and existential quantifiers give the logician mastery over multiple generality. Do all men love some women? If so, then $\forall x \exists y(x \text{ loves } y)$. Line up the men at the world's single's bar, and for each, there is some woman that he loves (generally not the one he is with, of course). For each man, that significant other may well be different, Philip remembering Irma with a smile of regret, even as Harry over there is busy reminding Daphne on the telephone that he loves her, and certainly he is still at the office. Every *man* is such that *he* loves *some* woman.

The positional reversal of existential and universal quantifiers yields a formula expressing the proposition that *some* woman is such that *she* is loved by *all* men: $\exists y \forall x(x \text{ loves } y)$. This is a very different business. Whatever declarations or reassurances that Philip, Harry, and the rest of the boys at the bar may be tendering to various dubious or disbelieving women, there is at least one woman, whether Mother Teresa, Helen of Troy, or Sophia Loren, such that each and every one of the boys loves *her*.

Frege's predicate calculus appeared on the screen of thought in the late nineteenth century, where it was viewed with some very modest appreciation by one or two logicians, the mathematicians, of course, too busy watching Poincaré and other big-time big shots on the screen of the adjacent multiplex to pay Frege much mind.

The predicate calculus survived their indifference; it is now the Universal Characteristic of mathematicians everywhere,

Leibniz and Frege having succeeded in imposing their vision of a symbolic world on everyone else.

THE QUICK AND THE DEAD

■ ■ ■ ■ I had taught mathematical logic at Stanford and at Rutgers and then again in Paris, and thereafter I found myself trapped in an academic wormhole that no matter what I did seemed inevitably to empty itself on the lawns of various sun-bleached California colleges. By late summer, when the academic year commenced, the lawns had all dried out, except for a few showpieces kept moist in front of administration buildings, and the great tawny light covered the campus in a pulsing glow, throwing fantastic blue shadows on the walkways between buildings and seeping into every classroom and library stack. My students were largely Vietnamese, and having made their way from Vietnam to California by way of the open South China Sea, they were not about to regard mathematical logic as anything more than a slight impediment on their otherwise straightforward climb from the bottom of the American social structure, where they worked harrowing hours at restaurants and car washes, to somewhere near the middle, where they pictured themselves as lawyers, occasionally as doctors, nurses, certified public accountants, computer technicians, and now and then, curiously enough, as politicians, the ordinary accoutrements of American life, such as microwave ovens, toasters, compact cars, stereo systems, and running shoes, which they acquired by means of the miracle of credit, coming to stand between them and the awful things they had left behind.

My colleagues, on the other hand, existed as a series of sharp caricatures, almost as if they had each read an academic novel before acquiring their own identity—the tall ungainly professor of statistics, forever trembling on the verge of a nervous breakdown and then toppling into frank hysteria at faculty meetings or upon reading some mean-spirited graffiti in the faculty washroom; the clever dialectician, clomping around campus in heavy steel-tipped work boots and Farmer Furd jeans, still arguing in favor of Stalin and the working class after all these years; and inevitably a number of mathematicians and physicists of real intellectual power who, because of fear or a sense that it just wasn't worth it, gave up their chance for the big time in favor of their own idiosyncratic pursuits, real estate, most often, an activity for which they were supremely unsuited, the moguls in the making that I knew all managing to lose money at a time when everybody else was making it.

In the fall of one gorgeous year, I taught the calculus; and in the spring of another gorgeous year, Gottlob Frege and I team-taught mathematical logic. Our classes were always well attended because logic was a prerequisite for an engineering degree, and they were, I must say, well received, Frege and I both receiving excellent if somewhat innocent standardized student evaluations, any number of students somehow saying the same thing, that while Mr. Berlinski should learn to match his ties and suits, *Mr. Frege is very nice.* No wonder they never complained about his clothes. Frege would dress severely, no matter the sunshine, which even in February seemed to light up every corner of the campus, wearing the same black frock coat and batwing collar that, no doubt, he had worn in Germany. You must imagine the man at the blackboard, the thick German chalk in his fingers, his back always toward our students and the logical symbols going up and down the board, the steps separated, when necessary, by heavily drawn lines.

We finished up our introduction to the predicate calculus in early March, and sometime just before the spring break, I found myself struggling to express the controlling insight of the course—and this book, of course—to students already symbol-punchy.

"The motion of the mind is conveyed along a cloud of meaning."

Meaning what?

I don't know, help me out on this.

"There is this paradox that we get to meaning only when we strip the meaning from symbols.... Right?"

My students look up, prepared to agree to absolutely anything, the brilliant, thrilling sunshine now occupying the whole of the classroom, encompassing my words in a burst of light.

And Frege? He stands there by the blackboard, fingering the chalk, but as is his habit, he says nothing whatsoever.

Absolutely nothing.

■ ■ ■ ■

RECONVEYANCE

The reconveyance of the propositional calculus as a formal system required a few modest adjustments in focal length, the system of symbols emerging clearly *as* a system of symbols after only a few turns of the microscope's threaded bezel. The predicate calculus is an incomparably richer symbolic system, and the machinery required to cast it in purely formal terms is correspondingly more detailed.

The formal system of the predicate calculus encompasses the whole of the propositional calculus, and I assume it is in place and that, towel on forearm, it is prepared to be of service. The primitive symbols of the predicate calculus go farther than the propositional calculus in comprising:

Individual variables:	$x, y, z, \ldots$
Predicate symbols:	F, G, H, ...

Symbols for various relationships: R, S, T, . . .

Symbols for quantification: $\forall$, $\exists$

In fact, only one quantifier need apply for the position as primitive; the two quantifiers are linked by definition, $\forall x Fx$ saying nothing more than $\sim\exists x\sim Fx$—if everything is F, nothing isn't, and vice versa.

A universe of symbols now occupies space in the spreading sunlight. Rules of grammar next act to specify the grammatical formulas and by so doing give the symbols their characteristic shape. Only two rules must be added to the rules already governing the propositional calculus. Boldface type again indicates that we logicians are talking *about* predicates (**F, G, H, . . .**), individual variables (***x, y, z, . . .***), or whole formulas (**A, B, C, . . .**):

1. If **F** is any predicate symbol, and ***x*** is an individual variable, then **F*x*** is well-formed.
2. If **A** is well-formed, and ***x*** is an individual variable, then $\forall$***x*A** is well-formed as well.

The rules prompt what I hope is the by-now familiar trudge up the inferential staircase, for they specify whether a formula is well-formed in terms of whether the formula occupying the step below is well-formed, and so engage the logician in his or her accustomed trek:

1. $\forall$xFx & $\forall$yGy	←	(Is this formula grammatical?)
2. $\forall$xFx	←	(Yes, if *this* formula is grammatical, and)
3. $\forall$yGy	←	(if *this* formula is grammatical.)
4. $\forall$xFx	←	(This formula is grammatical, if)
5. Fx	←	(this formula is grammatical, and)
6. $\forall$yGy	←	(this formula is grammatical, if)
7. Gy	←	(this formula is grammatical, but)
8. Fx	←	(this formula *is* grammatical.)
9. Gy	←	(Ditto.)

→ and so up the staircase to verify line 1 →

Grammar being dealt with and done, axioms and rules of inference need be added to the formation rules in order to make the system vibrate. The adumbration of the axioms requires—there is no help for it—a conceptual detour. Within the propositional calculus, the axioms make their appearance as *shapes of the very system*. A procedure of substitution, expressed by the rules of inference, then allows the logician to juggle like shapes so that he might see the soul of $(R \supset S) \supset (T \supset W)$ in the shell of $(P \supset Q)$.

The same procedure may be used in the predicate calculus, but it is complicated, tedious, and ugly. It is for this reason—plain laziness, too—that the logician repairs to *axiom schemata* instead of axioms when formalizing the predicate calculus. Axiom schemata do not themselves appear *in* the formal system. They are part of the logician's own vernacular, expressed in the same language that he or she employs to talk *about* formulas and predicate symbols. Each axiom schemata specifies the form of a formula, and each axiom of the system itself is obtained from the form as an instance.

Axiom schemata take infinitely many instances and this may well seem—and, in fact, it does seem—as if at a crucial moment the logician might be returning to precisely the concepts that lately he had abjured, rather like someone hoisting a glass at an AA meeting; but axiom schemata are conveniences, nothing more.

Or less.

Those axiom schemata, then.* There are only two that go beyond the clan of the propositional calculus. And they both require certain restrictions, which I have enclosed in brackets.

1. $\forall x\mathbf{A} \supset \mathbf{B}$ is an axiom—[*if* **B** is the very same formula as **A**, or *if* wherever **A** has a free

*This is in essence the system I learned from Alonzo Church many years ago.

occurrence of the variable x, then **B** has a free occurrence of some other variable y].
2. $\forall x(\mathbf{A} \supset \mathbf{B}) \supset (\mathbf{A} \supset \forall x\mathbf{B})$ is an axiom—[just so long as **A** does not contain a free occurrence of the variable x].*

Two rules of inference now; they are:

1. If $\mathbf{A} \supset \mathbf{B}$ is an axiom or a theorem, and so, too, **A**, then **B** is a theorem.
2. If **A** is an axiom or theorem, then so is $\forall x\mathbf{A}$.

The system is finished, and while it is always worthwhile to try to figure out what these strange symbols *mean* (and they do, of course, mean something), for purposes at hand it is also worthwhile to purge considerations of meaning from consciousness, the double maneuver of forgetting and remembering not only the prerogative of the logician but an indispensable human activity, one without which we would all be lost.

There remains the little matter of a demonstration, as car salesmen say. The logician wishes to establish that $\forall x\mathbf{A} \supset \exists x\mathbf{A}$. *Establish*—meaning prove absolutely. Stepping back for just a moment from the symbols themselves, we see that what this little affirmation actually affirms—and what little it affirms—is that if something is true of everything, then it is certainly true of something. It is hard to fault this claim as a claim, but our business is *proof,* and for that no feeling of contentment suffices.

The argument's taponade involves eight beats:

*The restrictions are explained more fully in the appendix to this chapter.

1. $\forall xA \supset A$	(axiom schema)
2. $\forall x{\sim}A \supset {\sim}A$	(ditto, with ~A instead of A)
3. $\forall x{\sim}A \supset {\sim}A \supset A \supset {\sim}\forall x\,{\sim}A$	(tautology)
4. $A \supset {\sim}\forall x\,{\sim}A$	(rule 1, lines 2 and 3)
5. $A \supset \exists xA$	(by definition with $\exists xA$ for $\sim\forall x\,{\sim}A$)
6. $\forall xA \supset A \supset A \supset \exists xA \supset \forall xA \supset \exists xA$	(tautology)
7. $A \supset \exists xA \supset \forall xA \supset \exists xA$	(rule 1, lines 1 and 6)
8. $\forall xA \supset \exists xA$*	(ditto, lines 5 and 7)

There is no denying that this argument, though trivial, is difficult to grasp. The symbols are alien and the steps undertaken stiff, so that the logician seems to be moving from assumptions to conclusions with very little suppleness. Nonetheless, the argument does proceed from step to step in a way that is entirely proof against doubt, each step following explicitly from the one before, with only explicit appeals allowed to explicitly formulated axiom schemata and explicitly formulated rules of inference. A machine could undertake the demonstration, printing lines and checking them, although it must always be remembered that whatever checking is being done is being done against a background of living human concerns.

Beyond the details of the inference, there is a style as well. When Frege was drawing attention to his magnificently detailed creation, mathematicians simply had no idea of the nuances demanded by formal precision. The style has become the common possession of computer programmers. It is still the subject of incomprehension and irritated indifference, and it is only the fact that these people seem to emerge from behind

*I have not completely punctuated these lines; the resulting slight ambiguity is easier to take than the requisite thicket of parentheses.

their computer consoles in order to roar off in their red Ferraris that persuades the rest of us that there may be something to all those details. But to appreciate the style itself, it is worthwhile to contrast the chain of inferences commanded by the logician with the back-and-forth common in other disciplines—the law, say.

And just how might attorneys interrogate the proposition that if everyone is blonde, someone is blonde as well? *Their* standard of proof is reasonable doubt.

> *The Court will now hear arguments in re* $\forall x\mathbf{B} \supset \exists x\mathbf{B}$.
>
> *Your Honor, at this time the defense would move for summary judgment.*
>
> *Duly noted. Overruled. Plaintiff may proceed.*
>
> *Plaintiff calls Miss Blonde World, 1995.*
>
> *Raise your hand, state your full name, do you so swear.*
>
> *I do.*
>
> *Miss Blonde World, for the record, can you tell us whether you're a blonde?*
>
> *Objection. Calls for expert testimony.*
>
> *Sustained.*
>
> *I'll rephrase. Have you ever been* called *a blonde?*
>
> *Objection. Speculative.*
>
> *Sustained.*
>
> *I'll withdraw. To the best of your knowledge, has anyone ever called you a blonde?*
>
> *Oh, lots of times.*
>
> *And for the record, you are some blonde.*
>
> *Immaterial. Defense is prepared to so stipulate.*

This is, of course, parody, but it is not parody by much, the lawyers' give-and-take absurd if only because it is addressed to questions that they cannot formulate and that they do not know how to answer. My point in introducing a parody is otherwise. The logician's style can very easily be mistaken for a form of fussiness. Taking it thus reflects an error in perception. The apparatus at his command allows the logician a rare degree of independence, power, and intellectual mobility, affording him in the case under consideration the ability to command the absolute absolutely in just eight lines, while the attorneys are yet busy wrangling over cases and citations, struggling to establish whether the very blonde Miss Blonde World is a blonde, and if so, to what end.

AND NOW THE TRUTH

The predicate calculus is an instrument of matchless subtlety, but like one of those watches that can simultaneously tell the time in four continents, determine the phases of the moon, and predict solar eclipses into the next millennium, the symbolic system that Frege fashioned, and that generations of patient logicians perfected, rather invites the sensible if skeptical question whether the system is really of any use. I mean use in a sense that is independent of the system's future incarnation in computer programming. An answer is forthcoming, plainly, but it depends, that answer, on the logician's characteristic maneuver in which symbols gain and shed their meaning with sometimes dizzying speed. Considered simply as a series of shapes, when $\forall x\mathbf{A} \supset \exists x\mathbf{A}$ appears at the conclusion of a proof, the conquest of doubt involved is somewhat suspicious because so far as these shapes go, doubt was never an issue. On the other hand, given their intended meaning, these shapes return with the message that whatever is true of everything is bound to be true of something, and this is news that while true, is hardly new.

I seem to have maneuvered myself between two forms of triviality. This is hardly an adequate answer to my own question.

The question invites a more serious answer, one that is now forthcoming. The proposition that $\forall x \mathbf{A} \supset \exists x \mathbf{A}$, when given its ordinary and intended meaning, is not only true but *logically* true as well. Logical truth has already made an appearance in the logician's tent in the form of a tautology; but in the case of the predicate calculus, the tent must be enlarged. The proposition that $\forall x \mathbf{A} \supset \exists x \mathbf{A}$ is *not* a tautology inasmuch as it has the plain propositional form of $P \supset Q$. It is, nonetheless, a logical truth.

It is this concept that must be defined. Now the predicate calculus hinges on the apparatus of variables, quantifiers, and predicates. The variables and quantifiers occupy themselves in a perennial tango; but the predicates do not vary nor do they fall under the sway of the quantifiers. They are placeholders, the B in Bx standing variously for being bright, being beautiful, being brazen, and so on, as before, to being blonde; the D in Dx standing for being docile, domestic, dutiful, doting, and so on, as before, to doggish. The logical truths are true—that much, for sure; they *remain* true whatever the interpretation of their predicates, so that $\forall x \mathrm{B}x \supset \exists x \mathrm{B}x$ is true whether B stands for being blonde or being brunette or for being an astrophysicist named Bertha.

The logical truths, logicians say, repairing to a much older vocabulary, are not only true but true in every possible world.

This is not a notion that may be found *within* the predicate calculus; but it is a notion that returns the predicate calculus to an older assortment of concepts. A *valid* argument is one in which *if* the premises are true the conclusion *must* be true. The nexus between premises and conclusion is precisely and unalterably the nexus of logical truth, so that *if* all men are mammals, and all mammals animals, *then* all men are animals, the hypothetical itself emerging as a logical truth and so something true in all possible worlds. It is this nexus that the definition of validity captures, removing some, but not all, of the mystery inherent in inference.

Having introduced the logical truths as somewhat slovenly strangers, the logician is naturally obliged to spruce them up. The tautologies required only an easy definition; not so the logical truths. To capture their content, the logician must involve himself in an act of multiple creation.

By a *world* (or a *domain*), he means a set of individuals or things—the set of all dogs or of all dishes or of all disasters (free association pairing Bowser, that tempting dinnerware, and the forthcoming disaster). Those worlds now enter into existence, like stars seen suddenly to shine.

The formula F*x* pairs an individual variable and a predicate symbol; with a domain fixed, both receive an interpretation, the individual variable, *x*, matched, say, to Bowser; the predicate, F, to the set of all individuals that just happen to be dogs. On this interpretation, F*x* simply says what experts might expect it to say: Bowser is a dog.

Interpretations are free for the asking. Some turn F*x* into something true; others do not. Assigning the distinguished physicist Murray to *x* induces the absurdity that Murray is a dog; the interpretation that assigns the physicists to F cancels the absurdity. Murray is a physicist.

Assignments are stroboscopic in nature: they light up the quantificational apparatus in a brief burst of meaning. The assignments that place individuals within the proper predicates—Bowser and the dogs, Murray and the physicists—are said to satisfy the formula. Or not: witness Murray as a dog, or Bowser as a physicist.

The double concepts of assignment and satisfaction suffice to define a concept of logical validity. The procedure once again involves an ascent on the inferential staircase. An assignment satisfies a formula of the form $\forall x Fx$ under just the expected conditions. It must satisfy Fx—step one; and it must continue to satisfy Fx for every value of x—step two. In a world of dogs (and nothing but), an assignment sending Bowser to the dogs satisfies Fx. Step One. In satisfying Fx, it satisfies $\forall x Fx$, as well, no matter which dog x may be. What else is there in *this* world besides dogs? Step Two.

An arbitrary formula of the predicate calculus is valid in a particular world if it is satisfied by every interpretation. In a world of dogs *and* physicists, $\forall x Fx$ is not valid. No matter the interpretation of the predicate letter, there is something loitering out there that tends to spoil the satisfaction. Say that F designates the physicists, and Bowser will not do; say that F designates the dogs, and Murray will not do.

But the formula that $\forall x Fx \supset \exists x Fx$ *is* valid in this very world, no matter the interpretation of F and even if F has gone to the dogs and x to a physicist. This formula is hypothetical, a horseshoe mediating between all Fs are x and some Fs are x. An assignment satisfying the antecedent must satisfy the consequent as well. The interpretation of x as Bowser and F as the congenial community of physicists fails to satisfy Fx and so fails to satisfy $\forall x Fx$. Leaving the antecedent unsatisfied, the assignment satisfies the formula as a whole. Hypotheticals are *true* if their antecedents are false.

And thus to all other assignments in a given world.

And thus finally to all other worlds. A formula of the predicate calculus is valid if it is true in every world.

The machinery of the predicate calculus is intended to capture the logical truths in its theorems and vice versa. Such, at least, is the hope. But it is not a hope that I have yet fulfilled. I have not said whether the predicate calculus is complete, or whether it is anything else, in fact.

Students in Calcutta or Shanghai, take note.

And the rest of us might think of this: wherever the predicate calculus may be going, the concept of an algorithm is just about to hop off.

CHRONICLE OF A DEATH FORETOLD

The predicate calculus is a formal system, one doubling during recess as a symbolic system. Capturing the connection between premises and conclusions of the valid arguments, the system seems at once to be about everything and nothing, truth in every possible world narrowing itself to propositions whose power is achieved by an evacuation of their content. Frege's concern was with arithmetic and whatever else it may be, arithmetic seems rich beyond triviality.

A productive paradox may now be seen in the history of thought. Arithmetic is where the content lies, and not logic; but logic prompts certainty, and not arithmetic. Concepts are sliding against one another, the friction thus engendered kindling a fitful glow on nineteenth-century mathematics, and, indeed, on late-nineteenth-century life.

We know what we know in terms of what we think we know, every claim to certainty resting on yet another claim to certainty, until the chain of claims simply stops, as with a system of axioms, or with things that we are simply disposed to accept without argument and so without reason.

It was Frege's secret ambition to snap this chain of contingencies by discovering within logic itself powers sufficient to encompass arithmetic. This is not simply an interesting idea: it is a dense, powerful, commanding idea, and, indeed, it is one of the great ideas, for it holds open the promise of anchoring one ancient system of human thought—arithmetic—in the bedrock of a place that requires no additional justification because it admits of no additional doubt. Frege published his masterpiece, *The Foundations of Arithmetic*, in 1884; within its pages, arithmetic finds expression in what Frege considered purely logical terms.

The story of Frege's undertaking has a peculiar literary quality, almost as if it were being told by one of the South American fabulists, so that *The Foundations of Arithmetic* appears in the history of thought as if it were a chronicle of a death foretold, something that I myself seem to have foretold as well.

■ ■ ■ ■ Traveling from Santiago to Buenos Aires, Don Pedro de Los Angeles carried with him a green parrot, a monkey with a ribbon tail, and a locked steamer trunk. His wife, the beautiful Señora Sabrina, whose green eyes were as deep as the sea, greeted the parrot and the monkey with cries of joy, but when she asked what was in the trunk, Don Pedro said nothing, shaking his head and giving orders to the servants to carry the trunk to a closet behind his study on the third floor. Days and years passed. Don Pedro's full black beard turned white and cataracts shrouded his eyes. He walked slowly with the aid of a cane. The beautiful Señora Sabrina had long since become stout, her flesh quivering as she walked, and the low lovely voice in which she had once sung songs of love grew hoarse with age. One day, Don Pedro fell ill with the ague and sensing that his end was near, he withdrew to his bedchamber on the third floor of the white villa with the turquoise shutters. He suffered for four days, but on the fifth, his mind was clear. After the servants had left, his wife approached his bed. "Don Pedro," she said, "I have never asked anything of you but the love to which I am entitled, but I wish to ask a favor."

Don Pedro said nothing.

"Don Pedro, for forty years I have wished to know what was in the trunk you brought with you from Santiago. Satisfy my curiosity now, for you know I will never look in the trunk without your permission."

"Señora Sabrina," said Don Pedro, "there is a manuscript within the trunk. It is bound in vellum. The sheets are written in parchment. It is very old, older than the dawn of time, and a copy of the manuscript survived the great fire that lit the skies of Egypt and consumed the libraries of Alexandria."

"A manuscript?" Señora Sabrina asked in astonishment. "All these years you have clutched a manuscript to your heart?"

"Yes," said Don Pedro.

"Does it contain secrets, Don Pedro?"

"I do not know. I have not read it. It is foretold that all who read it must go blind."

Señora Sabrina looked at her husband's sightless eyes without saying anything.

"But you must know something of this manuscript," Señora Sabrina cried out in vexation, a trickle of perspiration falling between her breasts like water sliding between two mountains.

"Mere possession of the manuscript is itself a blessing," said Don Pedro. "Did it not cure Ramón Fernández of unendurable melancholia many years ago?"

"That is all very well, but what does it *say*?"

"The manuscript contains a series of numbered propositions, written in a very careful hand. Each proposition is said to have the unique ability to express and to exhibit the truth, so that reading these propositions, a man would know where the jaguar goes at dawn, and what will be the date of his death, and why the whale cries in the sea at night."

"And do you know the date of your death, Don Pedro?"

"Yes."

That night Don Pedro died peacefully in his sleep. Señora Sabrina sat for two days by his body, as is the custom, and on the third day, she withdrew the key to the steamer trunk from the mahogany box in Don Pedro's writing desk. Carrying a candle, for the closet had no windows, she bent stiffly, blew the dust of years from the trunk's lid, and with trembling fingers turned the lock. A dark, rich smell emerged from the interior. Señora Sabrina brought the flickering candle close and peered inside.

There was nothing there.

■ ■ ■ ■

It was Frege's great ambition to show that contrary to every easy natural expectation, arithmetic shared in the certainty that pure logic afforded the logician, the cathedral of arithmetical inference in the end logical in nature and so given over to very simple sacraments.

The idea that arithmetic is a form of logic is by no means obvious, but as so often happens in the history of mathematics, Frege's thoughts were invigorated by thunder booming on nearby hills, the concept of a *set*, the subject of a new, far-fetched but far-ranging theory seeming somehow to combine certainty and logical content in a powerful formal structure.

THE TARTAR OF MATHEMATICIANS

Let me conjure up by purely verbal means and maneuvers, a selection from among the world's things—an apple, a pipe, a drooping rose. Thus conjured, these items detach themselves from their accustomed places (a bowl of fruit, a rack of pipes, a crystal vase) and, thus detached, admit of assimilation into a *set*—{apple, pipe, rose}, the brackets serving somehow to solidify an act of intellectual assimilation so primitive as to be a property of the race. It is profitless to ask what a set is or how the mind collects objects for assimilation;

the questions invite only the obvious answers: Who knows? and It does. If sets are too primitive to be defined, they can, at least, be distinguished from other things. They are not heaps, of course, for it makes no sense to talk of a heap of numbers, and yet the numbers from 1 through 5 form a perfectly acceptable set—{1, 2, 3, 4, 5}. Still less are heaps sets. A heap of sand does not yield a set of sand. The heaps belong off with the mounds, the piles, the stacks, the masses, and the batches; the sets are otherwise—remote, detached, abstract. Heapwise, a heap of sand is as heapish as it gets. There is no heap comprising the heap of sand itself. Not so, the sets. Set formation may be self-applied. The set consisting of *five* numbers {1, 2, 3, 4, 5} may itself be collected into a set—{{1, 2, 3, 4, 5}}, this set consisting of *one* object, the set {1, 2, 3, 4, 5}. The curious thing about sets is that while they serve to collect various objects, they serve simultaneously to enlarge and so to change the stock of objects on hand, an interesting example of a quantum effect in pure thought. This process may be continued indefinitely, the innocent processes of isolation and assimilation rocketing upward, with the result that the triplet of apple, pipe, and rose suddenly seems the source of infinitely many new objects: {apple, pipe, rose}, {{apple, pipe, rose}}, {{{apple, pipe, rose}}}, and so to the set-theoretic stratosphere.

The transformation of set theory from metaphysical pastime to mathematical pursuit was undertaken by Georg Cantor in the late nineteenth century, and unlike the dour Frege, who managed with great care to stoke and bank the fires of his creative genius, Cantor alternated between the rhapsody of mathematical creation and frank mental illness, withdrawing at strategic intervals to various sanatoriums.

Cantor was born in St. Petersburg in 1845, but when his father became ill, he found himself transplanted to the alien soil of Frankfurt, and even though he thoroughly assimilated himself into German culture, there remained in his personality something of the defiant exotic. His mathematical talent was evident early on; like all of the great mathematicians, he was able to pick out complicated melodies on some purely mental piano after only a single hearing. He was educated at the University of Zurich and then at the University of Berlin, where he came under the influence of Karl Weierstrass, Ernst Kummer, and Leopold Kronecker, eminences of the German mathematical establishment, the latter destined to torment him with skeptical doubts. For a time, Cantor taught at a private girls' school in Berlin. It is a scene too rich to be relegated to obscurity: Cantor himself, dressed soberly in black, a roomful of adolescent German *mädchens,* twittering as he entered the classroom. Thereafter, Cantor joined the faculty at the University of Halle, remaining in Saxony for the rest of his productive life.

Set theory is, of course, his great creation; the story of its discovery is rich and fascinating, but it is not *my* story—at least not in the context of this book—and I have space only to skim its surface.

Isolation and assimilation are the fundamental mental motions upon which set theory depends, isolation serving to pick out objects, things, or numbers, and assimilation, to collect them into something that in turn may serve as the object of isolation. Set theory engages itself with sets of numbers most often, and with sets available, the theory makes plausible a variety of combinatorial processes in which sets meet, merge, separate, turn on a dime, and otherwise do as they are told. Or as they wish.

The very most fundamental operation is set membership itself; the number 8 is a member of the set consisting of the numbers 2, 4, 6, and 8. Thus $8 \in \{2, 4, 6, 8\}$, even as Caesar $\in$ {the generals}, Marilyn $\in$ {the blondes}, and Don Pedro $\in$ {the Dead}.

Set theoretic inclusion places one set within another—{2, 4, 6} within {2, 4, 6, 8, 10} or $\{2, 4, 6\} \subseteq \{2, 4, 6, 8, 10\}$. But inclusion may be defined in terms of membership itself. The set of anacondas is included in the set of snakes. Whatever belongs to the set of anacondas belongs to the set of snakes. Inclusion has vanished in favor of set membership and a certain elegant displacement of attention.

The union of two sets $\{1, 2, 3\} \cup \{3, 4, 5, 6\}$ is culled from both sets: $\{1, 2, 3, 4, 5, 6\}$; their intersection, $\{1, 2, 3\} \cap \{3, 4, 5, 6\}$, from only the members common to both sets: just $\{3\}$, in this case. (The deceptive simplicity of these operations persuaded mathematicians in the late 1950s to attempt to teach set theory to toddlers, with predictably disastrous results.)

Within set theory proper, there are sets, and sets of sets, and sets of sets of sets, and at the opposite end of that ladder, the empty set $\{\ \}$, one containing nothing whatsoever, its brackets vibrating ominously around something like the physicist's false vacuum.

There are infinite sets, of course, like the set of all natural numbers, and then there are still-*larger* infinite sets, like the set of all real numbers.

■ ■ ■ ■ Still-larger infinite sets? I seem to remember making this comment to dinner companions in the south of France—a California attorney with dark eyes and an inconceivably rich tan, and his pretty wife, who was said to be an actress, and was forever pursing her lips or tucking her short hair behind her pretty seashell ears. We were dining at a restaurant called the Golden Sail. The sea was spread before us.

"Wait a minute," said the attorney. "Infinite is infinite, am I right?"

"Absolutely, you're right, honey," said his wife. "Absolutely."

I then proceeded to show the four of them—another lawyer was lurking in the background, together with *his* wife—Cantor's ingenious diagonal argument.

Two sets are the same size if they can be paired up, member by member. The even numbers and the odd numbers are similar in size by this definition, each even number matched to the odd number it succeeds, so that 2 is matched to 1; 4 to 3; and so on. Somehow, this seems as it should be.

But what of the real numbers? These are numbers that must be represented by their decimal expansions, as when π is expressed as 3.14159 . . . the dots following in 9's turbulent wake signifying that the expansion goes on forever.

Everyone had for the moment stopped eating, the occasion marking a dizzying zenith in my own capacity to command *anyone's* attention.

Suppose now that the real numbers are enumerated in a matrix. Read from left to right, the matrix captures the decimal expansion of a particular real number. The subscripts indicate both row and column. The first entry of the first decimal expansion of the first real number is a_{11}; the second, a_{12}, and the third, a_{13}, . . . For purposes of illustration, I have set π as the first real number in the matrix, $a_{11} = 1$; $a_{12} = 4$, $a_{13} = 1$, $a_{14} = 5$ . . .

π	3	1	4	1	5	9		. . .		
	R_2	a_{21}	a_{22}	a_{23}	a_{24}	a_{25}	a_{26}	. . .		
	R_3	a_{31}	a_{32}	a_{33}	a_{34}	a_{35}	a_{36}	. . .		
	R_4	a_{41}	a_{42}	a_{43}	a_{44}	a_{45}	a_{46}	. . .		
	R_5	a_{51}	a_{52}	a_{53}	a_{54}	a_{55}	a_{56}	. . .		
	R_6	a_{61}	a_{62}	a_{63}	a_{64}	a_{65}	a_{66}	. . .		
	⋮	⋮	⋮	⋮	⋮	⋮	⋮	. . .		
	R_n	a_{n1}	a_{n2}	a_{n3}	a_{n4}	a_{n5}	a_{n6}	a_{n7}	a_{n8}	. . .

"Now this list," I went on, as all around us the Mediterranean Sea bathed itself in light, "is supposed to contain *all* of the real numbers."

"Meaning?"—this from Coppertone, LL.B.

"Meaning that even though the list goes on and on, sooner or later one is going to encounter every single real number, even if one has to descend through ten billion places to find it. And if every real number is on the list, the real numbers can be enumerated."

"Meaning?"

"Meaning that they can be placed into correspondence with the natural numbers, so that we can talk of the first real number, the second real number, and the third real number."

"So?"

"So if we can place the real numbers into correspondence with the natural numbers, then they're the same size and all."

"But you can't, right?" said the attorney's wife, who had left off primping and was staring at the napkin on which I had written the matrix of real numbers.

"What do you mean, you can't?" said the attorney. "I mean he just did. Look at the list."

"What I mean is that no matter how large the list, I can always come up with a real number that isn't on the list at all."

"So you come up with some number, how do I know you're not going to find it ten years from now, somewhere down on the list?"

"Listen, honey."

"No, I'm listening. I mean, I want to know."

"You don't have to worry about what's going to happen in ten years. I'll show you a real number that simply can't be on the list."

"So show me."

"Honey."

"Look at the diagonal I've drawn through the matrix."

I tapped on the arrows with the tip of my fork, one after the other.

"What I'm going to do now, I'm going to give you a simple rule for generating a real number, one that involves just looking at the arrows. Are you with me?"

"So far."

"The first number is 1, the second a_{22}, the third, a_{33}, the fourth a_{44}, and so on down the matrix. The arrows make it pretty clear what's going on. So anyway, using this rule, I can construct the decimal expansion of some particular real number R: 1, a_{22}, a_{33}, a_{44}, . . ."

"This I understand, but how do you know that *this* number isn't on the list?"

"I don't."

"So what's the big deal? From where I'm sitting, we're where we were."

"Honey."

"No, no, no, that's fine, but what I'm going to do now, I'm going to construct a number that can't be on the list."

"So construct," said the attorney, his white teeth smiling against his tanned face.

"It's very simple. I add one to every number in the decimal expansion of this particular number. This gives me a brand-new different number R*: 1 + 1, a_{22} + 1, a_{33} + 1, a_{44} + 1 . . ."

"But isn't that just the same old, same old?"

"No, it isn't. This number can't be on the list."

"Why?"

"Because it's different from every number on the list. It's different from π because it's got 2 where π has 1, right? And it's different from R_2, because it has a_{22} + 1, where R_2 has only a_{22}, right?, and it's different from R_3, because, . . ."

I stopped talking. The attorney sat there, concentrating, his brow for the first time knotted.

Then I happened to look in his wife's eyes. They were alive with pleasure.

"That's beautiful," she said softly. "Beautiful, beautiful, beautiful."

■ ■ ■ ■

THE SHORT GOOD-BYE

The universe that Georg Cantor created was at once the universe he inhabited, his mind forever unfolding itself within the strange, oddly pullulating structures that he imagined and so brought into being—sets, sets of sets, infinite sets, still-larger infinite sets, the system starting from a seed and growing thereafter with all the energy of genius.

Whatever the strange structures that Cantor envisaged as his nurses draped cold sheets over his flushed forehead, the seed of the system seems to have a simplicity that is entirely pure and so entirely simple. That seed is simply the concept of a set itself. There is nothing more fundamental in terms of which it may be explained and so it seems to offer the logician an additional primitive concept in terms of which logical inference may be defined. Like the predicate calculus itself, set theory seems at once to be about everything in general and nothing in particular.

It was thus that Frege was tempted to a fateful step, endeavoring to assimilate arithmetic itself, and everything that followed from arithmetic, to a form of logic that included the elementary concepts of set theory. The details of Frege's laborious reconstruction of arithmetic along logical lines are not directly of concern to us, but the essential idea conveys the essence of the whole. The lock of arithmetic opens, of course, with the key of the natural numbers, and these simply go on and on. But so, too, the sets. The number 0 may thus be identified with the empty set {∅}, the number 1 with the set containing the empty set, {{∅}}, the number 2, with the set containing the set containing the empty set {{{∅}}}, and so farther along the endless chain of numbers, each number matched to a set capable of conveying precisely—and only—its position in that ascending chain, the numbers vanishing in a puff of smoke, arithmetic revealing itself at long last as a part of the logically irrefragable aspects of human thought.

When Leibniz imagined the universe coming into existence as the result of a play between 0 and 1, he might have taken the dream still farther. On Frege's scheme, the universe comes into existence as the result of the play between nothing and a single operation; the empty set and the process of set formation bring the grand cascade of numbers into their indentured existence.

That crack-up is coming; it has come. By 1893, Frege had expanded the ideas presented in *The Foundations of Arithmetic* in a volume entitled *The Basic Laws of Arithmetic*. In 1903, he was set to publish the second volume of *The Basic Laws of Arithmetic*. The book was already in press, when he received a letter from the young Bertrand Russell. The letter was courteous, respectful—and devastating. The very concept of a set, Russell revealed, was inconsistent. His argument was too simple to be ignored and sets were too fundamental to be replaced. Russell asked Frege, and by means of the foghorn of history, he is asking us, to consider sets that do not contain themselves as members. Such are the normal sets. The set of all dogs is not a dog, nor is the set of all numbers a number. On the other hand, there are sets that *do* contain themselves as members. So long as sets are among the things, the set of all things is again a thing. Such sets are *ab*normal.

What of the set of *all* normal sets. The normal sets include sets of dogs, sets of physicists, sets of blondes, sets of whatever, and the set comprising all of them is composed of sets of sets, these including at least:

{{dogs}, {physicists}, {blondes}, {and all the rest}}.

Call this set N.

Is N normal? This is Russell's question. If so, N may be found among the normal sets and N is at least:

N = {N, {dogs}, {physicists}, {blondes}, {and all the rest}}.

But if N pops up in N, N is a member of itself—just look!—and by definition abnormal. It follows that N is at most

N = {{dogs}, {physicists}, {blondes}, {and all the rest}}.

Now, by definition again, the normal sets comprise all sets that are not members of themselves. Absent N as a member of itself, and N turns out to be normal and not abnormal.

But N must be either normal or abnormal, yet by the swinging arguments given, it would appear to be both, or neither.

A paradox has been reached.

"Hardly anything more unwelcome can befall a scientific writer," Frege wrote in the appendix to his book, "than having the foundations of his edifice shaken after the work is finished. I have been placed in this position by a letter from Mr. Bertrand Russell. . . ."

DEATH COMES FOR THE LOGICIAN

Sometime in the fall after the spring in which Frege and I had taught logic together in California, my great friend, the logician DG took his life. He had loved someone a great deal and for a very long time, and when it was over he had only logic left and logic was not enough. He was cremated in

Colma at the insistence of his wife; I watched as the conveyer belt took his coffin toward the winking red lights; there was a roar from far away as the gas-fired jets ignited, and two hours later, I was given a plain wooden box with his ashes.

I took the box with me to one of those sparse California hills, which are covered with chaparral and a few scrub oaks standing in copses.

I was about to scatter the ashes when I noticed that Frege had joined me. He was dressed, as always, in black. I opened the box and let the salt-smelling wind carry the ashes far away.

Frege looked into the middle distance. I thought he would remain silent.

"I always come for my own," he said, just before vanishing himself, leaving me alone with the smell of wild sage.

■ ■ ■ ■

APPENDIX: SOME RESTRICTIONS MAY APPLY

The precise details of inference in the predicate calculus are apt to seem onerous. Two points deserve comment in this respect. First, the details, although onerous, are not difficult, requiring only paper, pencil, and patience for their mastery. The calculus, by way of contrast, contains concepts of very real intellectual difficulty. Second, the details are in service of a very powerful system of thought, inference in the predicate calculus representing a massive and not simply a modest enlargement of previous thought, with even the most sophisticated logical systems of Boole or Venn seeming very much like toy choo-choo trains in comparison to Frege's smoothly chugging, diesel-pulled midnight express.

Those details then: Recall the axiom schemata presented earlier. The first axiom schemata says that any instance of a formula of the form $\forall x\mathbf{A} \supset \mathbf{B}$ is an axiom with the following restrictions: Either (1) **B** is the very same formula as **A** or (2) wherever **A** has a free occurrence of the variable x, **B** has a free occurrence of some other variable y. The axiom $\forall x\mathbf{A} \supset \mathbf{B}$ is true enough if $x = x$, $\mathbf{B} = \mathbf{A}$, and **A** is Fx, whence $\forall x\mathrm{F}x \supset \mathrm{F}x$ is an instance. And true enough again if $x = x$, $\mathbf{A} = \exists y\mathrm{B}(x, y)$, and $\mathbf{B} = \exists y\mathrm{B}(x, y)$. Thus $\forall x\exists y$(x is blonder than y) $\supset \exists y$(x is blonder than y) is an instance of $\forall x\mathbf{A} \supset \mathbf{B}$, but so, too, $\forall x\exists y$(x is blonder than y) $\supset \exists y$(z is blonder than y), with z standing in for x in $\exists y$(x is blonder than y). The substitution of z for x remains within the compass of the restriction. The variable x is free in $\exists y\mathrm{B}(x, y)$, although bound in $\forall x\exists y\, \mathrm{B}(x, y)$, and the variable z is free in just the same place in $\exists y$(z is blonder than y).

The substitution of y for x, however, involves a violation of the restriction and leads to the absurdity that $\forall x\exists y$(x is blonder than y) $\supset \exists y$(y is blonder than y). No blonde is blonder than herself. Bound variables clash in this formula, as logicians say. The restriction is designed to prevent the clash.

Restrictions on the second axiom schemata follow the same pattern.

CHAPTER 5

Hilbert Takes Command

With the appearance of Russell's paradox, the Thing that had been waiting to strike struck. The very foundations of mathematics were infected, the attempt to derive arithmetic from an intuitive and plausible form of logic doomed. Russell communicated his paradox to Frege in 1903, and thereafter, the mathematicians of Europe may be seen lifting their hands to their foreheads in a gesture of delicate dismay. To everyone save for the disgruntled few who had mocked the subject from the start—Leopold Kronecker, most obviously, who may be heard snickering, even now—set theory seemed at once too simple and too deep a discipline to be discarded. Like electrons in particle physics, sets are simple structures, *in*definable in terms of anything simpler; like quantum electrodynamics, set theory represents a rich, intriguing, and beautiful body of insights, a very "paradise," in the ebullient words of one mathematician. But Russell's paradox belonged to a malignant pride of paradoxes, all of them similar in nature, and together they cut very deep into the mathematicians' collective confidence, the mathematicians learning the lesson that everyone must sooner or later learn, that what is simple is not necessarily safe and that what is beautiful is not always true.

CRISIS AND RECOVERY

Rueful, chastened, but combative, mathematicians concerned with certainty set about to repair the infected foundations of their subject during the first two decades of the twentieth century. On the continent, the Dutch mathematician Luitzen Brouwer reposed his confidence in a form of intuition, a faculty of mind contemplating the progression of the integers as they rolled forward. The very idea that mathematics required a foundation disappeared from intuitionism, replaced instead by the interplay between a mental searchlight and a mathematical series, an interplay that, needless to say, Brouwer could no more justify than explain.

The most sustained and intelligent reconstruction of mathematics was undertaken as a specific form of masonry. Studying the paradoxes, mathematicians realized that they seemed to involve self-reference, an old, troubling logical intruder known since the Greeks. The liar's paradox is an example; the liar who says, "I am lying," seeming to say what is true only if what he says is false and seeming to say what is false only if what he says is true. Russell's paradox also hinges on self-reference, making its entrance at the moment that something turns in on itself, the normal sets struggling to appear normal and finding themselves abnormal, and vice versa. This suggested to mathematicians that set theory might be rescued, if not redeemed, if the conceptual machinery of set theory were recast.

In 1908, the Swiss mathematician Ernst Zermelo published a new set of axioms for set theory. There are ten axioms in what has come to be called Zermelo-Fraenkel set theory (the Fraenkel, Abraham Fraenkel, who added to the formation of the axioms somewhat later). Some simply cover the ground covered in Cantor's naive set theory (a superb choice of words, that), specifying the ways in which sets may meet and merge, but others are very carefully crafted so as to make self-reference difficult, contained, constrained, or simply impossible. Critics observed that a foundational discipline in need of such support was rather like an athlete gamely waving crutches; but with those paradoxes in plain sight, many mathematicians were persuaded that crutches were better than nothing at all. To this day, no one quite knows whether Zermelo-Fraenkel set theory can do what it is intended to do, which is to evade the paradoxes; then, as now, the most one can say is, So far, so good.

Unwilling to abandon Frege's project, Bertrand Russell and Alfred North Whitehead undertook repairs along a slightly different line, creating what they called a ramified hierarchy in the three very large volumes of their *Principia Mathematica*. Sets go upward in that ramified hierarchy, getting bigger or otherwise fatter (sets of sets of sets...), but not downward; in this way Russell's paradox is canceled because to a large extent sets are forbidden to be members of themselves (as when the set of all normal sets squirms into itself). Russell and Whitehead *did* succeed in deriving arithmetic from logic and, by means of a promise of things to come, all of mathematics from the same source (the numbered definitions, theorems, and proofs persuading philosophers that what Russell and Whitehead had to say must be true because the *Principia Mathematica* was too long to be false). Some mathematicians agreed, but others observed that ramified set theory hardly has the power to commend itself as logically indubitable, the *Principia* prevailing only because Russell and Whitehead had by ramifying the hierarchy of sets, inadvertently lowered their standard of success.

But as so often happens in the history of thought, these reconstructive efforts were being superseded by events, even as they were being undertaken, the banging and hammering heard during the tense early years of the century drowned out by other sounds.

CAPTAIN AT THE BRIDGE

David Hilbert was one of the two great mathematicians whose careers straddled the nineteenth and twentieth centuries, Henri Poincaré, by common consent, the other. Hilbert was born in 1862, in Königsberg, the capital of what was then East Prussia, and the birthplace as well of Immanuel Kant, whose intellectual influence seemed sedately to radiate through the streets, the public squares, the university, and thereafter into the rolling countryside beyond (where in 1944 it was flattened by an advancing column of Russian tanks). Hilbert's ascent from the classroom to the bridge of mathematics as its captain had all of the calm, unhurried quality of the passage through the sea of an extremely large fish. From the first, he had the quality of command, his mathematical method and his mood imperial, so that by the time he had finished with a problem, he had not only settled the point at issue, but emptied the field in which it was embedded, leaving only a few scraps in his wake. And thereafter, he moved on, organizing or reorganizing one branch of mathematics after the other, casting lines of influence, displacing the attention of mathematicians, presiding over the entire scene, and generally sending frothy waves moving in every direction.

If achievement were any guide, Hilbert's name would be as well known as Einstein's, and among mathematicians, it is. There are infinite-dimensional Hilbert spaces in quantum physics, where they play an indispensable architectural role, and Hilbert bases, Hilbert invariants, and Hilbert integrals in mathematics itself. Hilbert revolutionized elementary geometry and unified analytic number theory, persuading mathematicians the world over that to their surprise, they were simply multiple copies of himself, thinking his thoughts, following his lead.

But Hilbert did more than control his time and place; like Einstein, he came to control the future as well.

At the Second International Congress of Mathematicians held in Paris in 1900, Hilbert presented the mathematical community with a list of twenty-three open problems. He was then in early middle age, a slight man of medium height, with thinning sandy hair. The dusty lecture room at the Sorbonne was stuffy. Hilbert spoke in German, but his remarks had been translated into French beforehand. "We hear within us the perpetual call," he said. "There is the problem. Seek its solution." The declarations, injunctions, questions, and exhortations go on and on, the calm, high voice roaming easily over almost all of what in 1900 was already a discipline almost too large to be encompassed by any one mathematician. Some questions that Hilbert posed were specific, hard as a dentist's drill. In asking mathematicians to find a proof of Riemann's hypothesis, he was asking mathematicians to do what they habitually did, and the very fact that no one has succeeded in yet doing it is evidence that Hilbert appreciated the problem's depth. But other questions were programmatic. When Hilbert asked mathematicians to address the general problem of boundary values, he was inviting them to reorganize territory that he could sense was ready to be reclaimed. The work might well involve a platoon of mathematicians, slapping one another's shoulders and boisterously getting on with it. And still other questions were cunning, Hilbert dividing the stream of time by means of diffident sly gestures, as when in his second question, he asked mathematicians to determine "the compatibility of various arithmetic axioms."

Their *compatibility*—meaning what? *Various* arithmetic axioms? Meaning what again? These questions were addressed to the future.

Small wonder that the mathematical community that Hilbert established in the small German town of Göttingen has become an object of retrospective veneration in the history of science, almost all of the great mathematicians of the twentieth century remembering (or wishing to have remembered) that once when they were young they saw the great man lecture, and afterward sat in the cafés and talked into the endless night.

HERR H.

If the optimism with which Hilbert faced the future was sufficient to convince his colleagues, looking backward, we can see that it must, that optimism, have been discharged so that Hilbert could convince *himself*. The drama of his own doubts and morbid discontents may be imagined playing itself out on the psychoanalyst's couch, rich material for the analyst's own analysis.

In cases presenting certain neurotic features, which we may call examples of hysteresis owing to their peculiar character of persistent alternation between two competing sources of psychic energy, the course of analysis is difficult, indeed, often problematic, for the analyst must not only be in a position to interpret the symptoms correctly, but he must as well assist the patient in resolving the contradiction that lies at the heart of his neuroses. In such cases, the patient appears to oscillate between states of mind reflecting the influence of his conflicting desires, often with great rapidity, so that the analyst often forms the mistaken impression that he is dealing with a form of mania, followed by depression, when, in fact, the symptoms reflect nothing more than the oscillation of the ego as it is moved to and fro by unconscious desires moving in opposite directions.

In the fall of 19—, I was asked to see Herr H., a member of a Prussian family who had achieved a certain measure of success studying difficult scientific questions. Herr H. was of slight build and medium height, with a high forehead, thinning hair, and an air of studiousness; he appeared somewhat nervous, speaking at first in a diffident tone of voice, an attitude which changed characteristically over the course of his analysis as he became by turns garrulous, demanding, defensive, and then tentative and uncertain.

By all accounts his childhood was uneventful. His father, Herr O., was a country judge, and, a man exemplifying the Prussian virtues of thrift, duty, discipline, and obedience—virtues which analysis has revealed are associated with a definite stage of psychosomatic development, a stage, we may permit ourselves to speculate, that occasionally comes to characterize not only an individual, but an entire culture as well. His mother, M.T., was considered an unusual woman, having an interest in philosophy, music, and numerology. As so often happens in such cases, Herr H.'s mother sought to compensate for the father's aloofness and distance by an exceptional sense of devotion to her son, going so far as to write his essays for school. Quite obviously, Herr H. remained the focus of his mother's attention, his sister E., whom he mentioned only briefly, seeming to have been virtually invisible in the context of the H. household.

Herr H. received the traditional education given to children of the Prussian middle class, passing through elementary school and gymnasium in turn. Curiously enough, Herr H. recalled his own educational achievements in preparatory school and gymnasium as distinctly mediocre, even though he received excellent marks in all subjects; he was, he recounted, in the habit of refusing instruction in the sense that he would not accept a new idea or principle until he had quite independently worked the details out in his own mind, a trait that his teachers attributed to a certain perverse obstinacy, but that analysts will recognize as an early rejection of the influence of the overbearing father. Nonetheless, Herr H. recalled with some pride, when he had worked out the details of a difficult problem, his mastery of the problem exceeded that of his teachers, facts which again must be recalled in the context of his family relationships.

Sometime early in adolescence Herr H. recognized what he described as a "calling." It proved difficult to determine whether this memory was an accurate reflection of his state of mind or whether it was a later construct, imposed upon an older memory which had for unknown psychic reasons been displaced; but in the course of analysis Herr H. revealed that he was aware of his calling from a far earlier age and, indeed, deliberately neglected certain of his studies owing to his belief that he would later come to achieve domination over his discipline on his own, an interesting example of an immature delusional state.

Herr H. complained initially of sleeplessness and a restless anxiety that he had suffered intermittently for years, a combination that analysts well know is particularly difficult to resolve inasmuch as insomnia itself produces anxiety which in turn produces insomnia. Despite his relatively meek and unprepossessing appearance, over the brief course of his analysis, Herr H. revealed himself to suffer from certain familiar personality distortions consistent with his childhood calling, falling under the general heading of *folies des grandeurs*.

Some years before the commencement of his analysis, Herr H. had happily married and had become in turn the father of a single child, a boy. Early in his career, Herr H. had solved an important problem, discovering, as he revealed in the course of analysis, that the solution of the problem must exist rather than actually demonstrating the solution itself. Herr H. referred to his discovery as an existence proof, and was somewhat taken aback and, indeed, became quite agitated, when it was pointed out to him that existence rarely requires proof to assert itself, Herr H. obviously sensing in my reproach criticism from the distant father whose own words were, as Herr H. acknowledged, "few and far between."

Despite the apparent tranquillity of his domestic and professional life, Herr H. had for many years suffered from the influence of two competing psychic urges, which he himself described as a quest for certainty and a concomitant and deeply dreaded sense of doubt. As one might expect, feelings associated with certainty were invariably the source of intense but temporary euphoria, which insidiously, as he described it, would be undermined by an intimation of doubt, which inevitably grew until it took possession of his faculties as a form of depression.

His mental life thus alternated between euphoria and melancholy, with the added clinical feature that both states of mind had fixed manifest but undisclosed latent contents. The fact that both certainty and doubt had obvious psychosexual components, arising from the euphoria of potency and the fear of impotence stemming ultimately from the castration complex that he had acquired owing to his father's stern and distant behavior, Herr H. rejected with great indignation, a reaction all too common among highly educated scholars.

Herr H. reported on a recurring dream in which he would after a long and arduous journey through a heavily forested countryside arrive at the gates of a splendid villa constructed in the Italian fashion. The sight of the villa in his dream brought about feelings of deep satisfaction, and a sense that his long quest was over. He would walk through the elaborate gardens of the villa with immense pleasure, but gradually as the dream progressed he would have the uneasy feeling that he was not quite alone, and that the villa was meant for someone else. The feelings of anxiety increased as inevitably his steps took him to the portico of the villa itself. Then he experienced a feeling which he expressed as "the same dread." He would be conscious of something behind him and for long moments be unable to turn to look at it. When he finally summoned the courage to turn around, he saw with a shudder that hanging by the neck from a prominent yew tree on the villa's grounds was the body of a man. Descending back down the steps, Herr H. would approach the corpse, whereupon he discovered with a feeling of horror that the hanged man was none other than himself. He observed with particular repugnance that the lenses in his glasses had both been smashed.

Shortly after reporting this dream, Herr H. terminated his analysis, despite my protestations. In such cases, the analyst must learn humbly to accept defeat, acknowledging the existence of symptoms that owing to their inherently conflicting nature are clinically intractable.

■ ■ ■ ■

THE PROGRAM OF HERR H.

God may well have made the universe by means of mathematical laws, and he presumably found himself spared doubt as, muttering with awful grim concentration, he got on with the work at hand, scattering space dust to the winds and allowing the galaxies to blossom in the night. Mathematicians are men and women, chained by circumstances and the contingencies of their talent. For the most part, their work is mundane. A conjecture is tendered. And settled. It occurs to someone to ask whether, and if so, how? A part of the immemorial edifice is put in place; another part discarded. The sun hoists itself in the sky; the repairman comes to fix the leaking toilet, the neighbor's dog runs after a car. And then it rains. It is at night, that the great, hairless Thing appears.

Mathematicians are as adept as anyone else at hiding themselves from themselves, and that sense of dread, which is, of course, nothing more than the flipped coin of joyousness, tends to disappear with the dawn. But when mathematicians do engage themselves with doubt, they must undertake the difficult psychological task of allowing the self to become the subject of itself.

The attempt to anchor mathematics in logical certainty had by 1918 ended inconclusively, and although mathematicians thought they had a form of set theory that worked, they did not know that it worked, and what is worse, they could not appeal to the Zermelo-Fraenkel axioms with an honest sense that here at last was the place they were prepared to repose certainty.

Working with other mathematicians but dominating their imagination as well, Hilbert refused to take part in various reconstructive schemes, attempting, instead, to achieve a vantage point that swept up and subsumed all partial efforts at repair and reconstruction. Such is the ambition that by 1918 came to be known as the Hilbert program, the eponymous nature of the program carrying a faint warning squeak that if Hilbert was prepared to live by the precepts of his program, he would have to be prepared to die by them as well.

Hilbert's program is as much psychological as mathematical and so involves a way of seeing and the achievement of a multiple perspective in which the mathematician contrasts what he sees with what he sees as he sees. In literature, art, mathematics, and life, the eye is our indispensable metaphor. Now in the normal course of doing mathematics, the mathematician sees the familiar objects of his speciality: numbers, groups, rings, the calculus, differential equations. Stepping back, he sees himself *proving* things; and it is an absolute commitment to proof that distinguishes the mathematician from mystics, psychologists, great masters of the law, physicists, chemists, biologists, writers, sculptors, moguls, maniacs, indeed, from everyone and everybody else. Proof is the mathematician's coin, the one that he must forever clutch.

Any particular proof is a part of a system of proofs, the system very often remaining out of sight as the mathematician goes about business. But all mathematical systems share a common structure. They depend on a certain stock of symbols (*what else?*), certain ways of combining those symbols (*without which, gibberish*), certain assumptions (*without which, nothing*), and certain rules of inference (*without which, chaos*).

Mathematical systems become formal when the logician imposes constraints on the loose-and-easy structure of ordinary mathematical practice, just as a promise becomes formal when the lawyer freezes the give-and-take of ordinary life into a contract. Formal systems require an explicit finite list of primitive symbols; an explicit and finite list of rules determining which formulas are grammatical; an explicit and finite set of axioms or axiom schemata; and explicit rules of inference governing the steps the mathematician may or may not make. No guesswork, intuition, or hunches allowed. The system is *mechanical* in nature, although like any machine—*any* machine—infused with the intelligence of its creator.

Above all else, formal systems require that peculiarly logical act of cunning renunciation in which the logician simultaneously evacuates meaning from the symbols of a system while retaining at some level of consciousness a firm understanding of what those symbols mean.

It is this act of renunciation that prompted Hilbert to a shrewd psychological maneuver. There is the mathematician's anxious night chatter, interrupted now by the calm, neutral voice of the analyst.

> *With meaning removed from symbols, what is left?*
> "The symbols themselves, the moves that they can make, the game that they play."
> *To what end, this game?*
> "To what end, any game?"
> *But what does a game have to do with doubt?*

Whereupon Hilbert responds: "Everything," having evidently learned something from his analysis all along.

If—*Yes? If what?* If one—if *we*—could establish something about the game, then doubt might be denied. The first step has been undertaken. The game has now been seen for what it has been all along. And just as suddenly seeing a woman's face emerge from a delicate network of black brushstrokes brings a new object into creation, seeing the game for what it is brings a new object of contemplation into the mathematician's universe. Where before, there were numbers, equations, groups, sets, fields, rings, and strange topologies, now there is, in addition, the game itself.

But to see something, it is necessary to see it from some place, the very act of seeing displacing the mathematician from his accustomed position *within* the game; displaced, and so somewhat discomfited, he sees the game for what it is, sees the symbols for what they mean, and sees all this in the familiar universe in which he can do anything, except, of course, catch himself seeing what he has seen.

The arena of the game is ordinary mathematics; but the vantage point, that is *meta*mathematics, and the distinction between the two, which was uniquely Hilbert's, organized for the first time those confused and anxious logical voices into a coherent whole. The fulcrum of certainty rests on a distinction. Mathematics is a game of symbols. And metamathematics is where the game gains meaning.

Confronting the chaos of the years just passed, Hilbert determined—indeed, he proclaimed—that mathematics finds refuge from doubt in metamathematics, in what the mathematician can prove, demonstrate, or otherwise indicate *about* that fabulous game of symbols that he has brought into existence and that he means now to certify and redeem.

ARITHMETIC REVISITED

The Peano axioms express the properties of the natural numbers (0, 1, 2, 3,...); but Peano arithmetic does not comprise a formal system, if only because Peano himself is perfectly prepared to stand hunched over various shoulders and, speaking in his lisping Italian, explain his own axioms in the language of ordinary mathematics. In formal arithmetic, the shell of the Peano axioms remains as a set of symbols, but along with the meaning of those symbols, Peano has been allowed decently to disappear.

Formal arithmetic envelops the predicate calculus and its rules and symbols, going beyond them in a number of ways. Five new symbols appear in its primitive vocabulary:

$$=, S, +, \times, 0,$$

the symbol "=" (for equality); another symbol "S" (for succession); another "+" (for addition); another "×" (for multiplication); and still another "0" (for zero). Within formal arithmetic, these five symbols function as shapes; my parenthetical explanation of their meaning comes from a perspective beyond formal arithmetic.

The predicate calculus says nothing about equality or its sign, "=", and so a few additional axioms must be added to the system already in place specifically designed to control the behavior of this symbolic particle. These axioms do precisely what they are designed to do. They offer no surprises. Assume them given. The result is the predicate calculus with equality.

The purely arithmetical axioms now:

1. $\forall x \sim (Sx = 0)$
2. $\forall x \forall y((Sx = Sy) \supset x = y)$
3. $\forall x(x + 0 = x)$
   $\forall x \forall y(x + Sy = S(x + y))$
4. $\forall x(x \times 0 = 0)$
   $\forall x \forall y(x \times Sy = (x \times y) + x)$

These axioms *say*—but, of course, *they* do not say anything at all. The logician does the saying; the symbols simply sit there, taking up space on the printed page. What the logician says is straightforward and hardly any different from what Peano might have said, indeed, from what Peano did say. The first axiom? There is no number whose successor is 0. The second? Two numbers that have the same successor are equal. The third and fourth? Addition and multiplication on the inferential staircase.

Beyond these four axioms, an additional rule of inference is needed, one intended to capture within formal arithmetic the principle of mathematical induction. Now in ordinary arithmetic, mathematical induction functions as yet another inferential staircase. *If* 0 has some property, and *if* whenever some arbitrary number has that property, its successor has that property as well, *then* all the numbers have that property. Starting from a specific place, a row of upright black dominoes extends far into the farthest reaches of space, along with the table on which they have been carefully mounted. Surrendering to temptation, an enterprising adolescent gives the first domino a tap. Down it goes with a lustrous click. What of the others? Are they, too, destined to fall in time? *Yes,* if whatever the domino and wherever its place, if it topples, so, too, its nearest neighbor.

In the Peano axioms, the principle appears as a separate axiom. Not here. Mathematical induction enters into affairs from above, as a command about formal arithmetic, a rule governing inference. The ordinary and familiar apparatus of the predicate calculus is pressed into service, the formula **A**(0) saying simply that 0 has some property that the predicate **A** expresses. That rule of inference, now*:

3. If **A**(0) and if $\forall x(\mathbf{A}(x) \supset \mathbf{A}(Sx))$ are theorems, then so, too, $\forall x\mathbf{A}(x)$.

The back translation from symbols: *If* 0 has some property, and *if* whenever some arbitrary number has that property, its successor has that property as well, *then* all the numbers have that property.

Nothing more. Formal arithmetic has acquired the mysterious capacity to whir and hiss and engage itself in inference. The inferences of ordinary arithmetic, which for centuries had lain beneath the folds of conscious thought, now emerge into the sunlight, blinking weakly, their bony hands raised high. Adding 1 to any number, for example, gets you nothing more than the number's successor. In symbols: $\forall x(x + 1 = S(x))$. Two hundred and thirty-one plus 1 is the successor of 231—232, as it happens. And the same for any number. The proof is entirely mechanical. The machine commences with a definition: 1 is the successor of 0. And thereafter, it writes the following four lines:

1. $\forall x(x + S(0) = S(x + 0))$
2. $\forall x(x + 0 = x)$
3. $\forall x(x + S(0) = S(x))$
4. $\forall x(x + 1 = S(x))$

*The first and second rules of inference have been given in chapter 3, recall.

ANNA OF ARITHMETIC

Reading a novel with an innocent eye, students very often lose themselves in its pages, making their decision about the novel on the basis of whether they felt comfortable or at home within its world and more often than not identifying the author with his or her protagonist, every novelist receiving from time to time letters addressed to his creation—*Dear Anna, don't do it.* Such is the triumph of art. But such is the triumph of illusion, as well.

After some experience, the student learns to step back, recognizing that *Anna, she's got to do what she's got to do,* and this because what she's got to do is artistically required. No one reading *Anna Karenina* is quite prepared to see her departing the novel, therapist in hand, and briskly getting her life together. A sense of literary sophistication begins when aesthetic standards are substituted for moral judgments. This makes art a profoundly amoral undertaking, but a profoundly interesting one as well.

Mathematics is, among other things, a form of art. Before Hilbert, mathematicians and logicians had banged around within the confines of various mathematical systems, hoping against hope to arrange the system so that it seemed entirely secure, the effort as doomed as the correlative effort to persuade Anna Karenina to undertake therapy.

Hilbert persuaded everyone to step back. Stepping back, mathematicians saw mathematics for what it might be, a formal game, the perspective cold but liberating. Thus removed from what they habitually did, mathematicians, like students of literature, were forced to ask not whether the Anna of arithmetic seemed nice, friendly, kind of snooty, confused, or otherwise irritating, but whether she made artistic or mathematical sense. A question of judgment had come to replace a question of certainty.

And with judgments come standards. They must, those standards, be chosen so as to reflect the original impulse yielding the decision to distinguish mathematics from metamathematics. And they must, those standards, be standards that can be met by proof, even if it is proof delivered in the metalanguage itself, for without proof, there is simply no mathematics at all.

Hilbert's standards of satisfaction were simple, straightforward, radical, and audacious. A formal system in general, and formal arithmetic in particular, must in the first instance be *consistent*. Without consistency, the very game of mathematics loses its point. The machinery of proof cannot grind on to produce $2 + 2 = 4$, while grinding farther to produce $2 + 2 = 5$. The mathematician must be in a position to establish that nowhere in the infinity of theorems generated by the formal system is he or she likely to find a statement of the form P and another of the form $\sim P$. The standard of consistency prevents symbols from appearing in forbidden juxtaposition. Plainly if certain symbols are forbidden to juxtapose themselves, no contradiction can arise, because the system simply lacks the power to express the contradiction.

The formal system must in the second place be *complete*. And this, too, is a metamathematical demand, a command issued by the symbol master. Formal arithmetic is intended to capture everything that is true of arithmetic and nothing more. The game loses its point if what is true about arithmetic and what can be demonstrated within formal arithmetic fail to coincide.

And finally, Hilbert urged, the formal system must be *decidable*. A finite and mechanical procedure must exist that determines for each and every claim made by the formal system whether that claim may be proved within the formal system.

A system that is consistent, complete, and decidable is proof against doubt. Witness the propositional calculus.

There is one more point, the last. Whatever the mathematician proves about the formal system must be proved using tools that are essentially no stronger than the tools found within the formal system. Otherwise the enterprise would once again have no point, doubt creeping in at the metamathematical level just as it was banished at the mathematical level.

And so almost to the advent of the algorithm, for Hilbert was asking for nothing less than the subordination of the whole of mathematics, with its far-flung and subtle concepts, to a mechanical routine—mechanical in its formation rules and rules of inference, mechanical in the verification of its proofs, mechanical in its ability to decide mathematical questions without thought, intuition, meaning, or deliberation. Mechanical as in a machine.

And mechanical, let me add at once, in a way that seems quite inhuman.

DISORDER AND THE DARK DEFILE

Hilbert pursued his metamathematical agenda with what one logician described as "all of his authority as a great mathematician." By 1930, he had every reason to believe that the community of logicians was completing the task that he had set for them. Logicians had obtained a number of partial results; Presburger, in particular, had demonstrated by impeccable metamathematical means, that a weak version of arithmetic (addition, but not multiplication) was indeed consistent and complete.

Returning to his birthplace at Königsberg in order to deliver a lecture, Hilbert closed his remarks with words that were later inscribed on his tombstone: "*Wir müssen wissen. Wir werden wissen.*"

We must know. We will know.

We realize now that that was the last time those words could have been uttered without irony.

Hilbert lived on as the magnificent institution he had created was emptied of its Jewish scholars by scrupulous Nazi bureaucrats. He became old as, all around him, a monstrous political regime stretched itself in the sun, its lizard tongue flicking. He continued to work, guiding a handful of students to their doctorates, but plainly, the enormous sense of command that had sustained him for years and then decades had slipped from his grasp. He tottered when he walked, and although he was sharp at times, at other times, he seemed to wander. Schiller's poems were read to him at night; he sipped at soup.

And then in 1943, after suffering a fall, he died in Germany alone; the mathematicians who had heard his voice and fallen under his command had scattered, some going to the United States or South America or even China, others, for all their sophisticated and intellectual cunning, finding themselves packed in freight cars, grinding their way to some place in the east.

CHAPTER 6

Gödel in Vienna

Having left Paris in order to leave Paris, I lived in Vienna in the fall of one year, and left in the spring of another. I seem to have liked leaving places. But while I was there, I liked Vienna well enough. It was dark brown and sepia and sooty and the waiters seemed to glide from table to table in the cafés at night. I lived in an apartment on the Oberdonaustrasse and from my window I could see the steeples of various baroque churches, poking into the morning mist or into the twilight. The ghosts were there to keep me company, sitting in the cafés or strolling grandly along the Kärntnerstrasse. Once I saw the Crown Prince Rudolf hurrying away from court in a fiacre, his coachman Bratfisch whistling, "*Wo bleibt die alte Zeit?*" while somewhere to the east of the city, in a country castle, his mistress waited, sloe-eyed, peach-lipped, a single rose pinned to her auburn hair. Time has now rolled backward, the city belonging to others, only the ghosts hanging around and waiting for the wheel to turn. It is now 1931. And I suppose that I am there waiting with the other ghosts.

Seven years younger than the century, the Viennese mathematician Kurt Gödel is a man of medium height, slight in build, with a fine symmetrical face, his hair slicked backward in the central European style. He wears tortoiseshell glasses, which enlarge his eyes. He has already established his reputation by demonstrating in his doctoral dissertation at the University of Vienna that the predicate calculus—the system of inference that Frege first devised—is complete. By means of the calculus it is possible to demonstrate every logically valid statement and only the logically valid statements. This is inspiring proof that by the judicious selection of axioms and rules of inference, it is possible to construct a formal system that captures, and captures completely, intuitions about the commanding concepts of logic and inference. The proof functions as a vindication of Hilbert's program.

Time inches forward. Gödel makes an unobtrusive appearance at a number of mathematical and philosophical seminars; he is a presence at meetings of the Vienna circle, where he says little but offers the ineffable impression of intellectual superiority; he is unfailingly helpful and polite to other mathematicians, who many years later will remark on his extraordinary ability to evaluate a mathematical problem quickly and clearly and indicate at once the main lines of its solution. He is reserved but not unsociable, poised and yet curiously aloof. He seems to have appreciated women, taking a number of them to the opera or to the symphony. His reputation is that of a young man with fine calm analytical abilities and a sly, vividly controlled sensuality. He is obviously high-strung, nervous, and often anxious.

In the autumn of 1931, Gödel published a paper of twenty-five pages with the title "On Formally Undecidable Propositions of *Principia Mathematica* and Related Systems." The reference to Russell's *Principia* is, in fact, a diversion; Gödel's paper is really about *any* axiomatic system in which the natural numbers may be described. It is thus a paper concerned with the oldest of mathematical ideas, the system of whole numbers.

Within the compass of those twenty-five pages, Gödel established that arithmetic is *in*complete, the Hilbert program doomed. What is more, he demonstrated that the consistency of arithmetic cannot be demonstrated by means of reasoning that is as simple as arithmetic itself. Freedom from contradiction is purchased only by systems whose own freedom from contradiction is problematic.

And in proving this, he also brought about the advent of the algorithm, giving, for the first time, a precise mathematical description of an old but hidden idea.

PROOF AND PARADOX

Gödel's theorem is based on a paradox, one first discovered by the French mathematician Jules Richard in 1905. Like Russell's paradox, this one turns on self-reference, immolating itself on itself. Ordinary English is often used to say ordinary things about ordinary numbers—that the number x is prime; that x is even; that x is larger than y; that x is unlucky or y lucky. These expressions define or specify properties of the natural numbers. Since expressions are made of words, and the words in turn of letters, there is no difficulty in arranging these expressions in a list. The list is infinite, of course: there is no end of things to say about the numbers, but every expression on the list returns itself to the finite letters of the English alphabet.

Suppose, then, that these defining expressions are listed simply: E(1) is the first, E(2) is the second, E(3) is the third, and E(n) is the nth, whatever number n may be. Nothing is hidden by the list. It is what it seems.

Now pick any natural number—27, say—and any expression on the list—the third, say—and suppose that E(3) says that x is a prime number. There are two possibilities: E(3) *is true* if $x = 27$, or not. In this case, not. The number 27 is not prime.

The argument now gathers momentum. Suppose that for any natural number n, the logician asks whether the nth expression on the list is true of n. Say that n is again 27. The twenty-seventh expression on the list is E(27). It says, I am assuming, that x is a multiple of 3. Just so. The number 27 is a multiple of 3. E(27) is true of 27. With other numbers, the situation is different. Say that n is 14 and that the fourteenth expression on the list is true of a number just in case that number is odd. E(14) is not true of 14.

Consider next a brand-new property of the natural numbers, one true of any number n just in case E(n) is not true of n. This is a perfectly reasonable property of the natural numbers, if somewhat baroque; it does nothing more than pick out the natural numbers that are not satisfied by their correlative expressions. The expression designating that property must appear on the list, somewhere farther down, perhaps among the very large numbers. Let us say that its position is q, E(q) saying that whatever the number n, E(n) is not true of n.

A paradox is now in prospect, and like all paradoxes, this one has two hands. Suppose that $n = q$. On the one hand, assume that E(q) *is* true of q. It follows that E(n) is *not* true of n, which in this case means that E(q) is *not* true of q. On the other hand, assume that E(q) is *not* true of q. Then E(n) must be true of n, which means that E(q) *is* true of q. In short, the number n is satisfied by the nth expression on the list if and only if it is not satisfied by the nth expression on the list.

It would seem that the intellect has run itself into a wall, using nothing more than the letters of the English alphabet, a list of sorts, and a fatal romance with self-reference.

THE CODE

It is the concept of proof that lies at the heart of Gödel's theorem, and like the argument just given, Gödel's argument involves self-reference, a sentence appearing from god-knows-where and saying of itself:

> I cannot be demonstrated.

A paradox would seem to follow at once. If this sentence can be demonstrated, then what it *says* is false, in which case it *is* indemonstrable. On the other hand, if this sentence cannot be demonstrated, then what it says is true, and the fact that it is indemonstrable indicates that it can be demonstrated. After all, the last sentence that I have written provides precisely such a demonstration.

But with truth and provability sunk in paradox, there yet remains a sentence on the margins of the meaningless that succeeds in referring to itself without crushing itself into absurdity:

> I cannot be demonstrated within a particular formal system.

This sentence may well be indemonstrable within the formal system and yet true enough, the demonstration of its truth taking place *beyond* the formal system, in the high country of metamathematics.

No paradox, surprisingly enough. There is no reason why the statement cannot be true, for in sober fact, it *is* true. Truth remains inviolable. Not so proof.

Gödel's theorem is about formal arithmetic, and formal arithmetic is simply a logically concise, wonderfully detailed way of packaging the axioms of arithmetic so that the skeleton of inference is by a process of imaginatively achieved translucence laid bare. There are axioms and theorems, as in ordinary arithmetic, but the theorems follow from the axioms by an appeal to rules of inference, and the symbols are specified with unusual care. Meaning has been evacuated from the system, the word *formal* signifying, as it has signified all along, that the logician has undertaken the by now habitual exercise in distance and double thought.

But if meaning can be withdrawn from formal arithmetic, it may be reintroduced as well when the symbols of the system are systematically invigorated by the logician's understanding. Withdrawal and reinvigoration constitute a back-and-forth process. In this way, the familiar theorem, that

$$1 = S(0),$$

which means, of course, that the number 1 succeeds the number 0, loses its arithmetical meaning in formal arithmetic and acquires instead a primary interpretation as a sequence of six shapes.

$$\underline{1 = S(0),}$$

the line underneath the symbols serving to call attention to the symbols themselves, *the very shapes,* rather than their meaning.

With a tilt of the head, the same sequence of six shapes reacquires its normal and intended meaning as a statement of arithmetic. The number 1—the *number,* note, not the numeral, which is merely a mark—is in fact the successor to the number 0.

If the logician has the capacity to infuse and then evacuate symbols of their meaning, it is a capacity he or she exercises from a perspective that is broad enough to encompass both formal and ordinary arithmetic. When he sees the symbols and she plays with their meaning, the seeing and the playing both take place from a metamathematical position. In *remarking* that $\underline{1 = S(0)}$ are the symbols or the shapes that *express* what

1 = S(0) means, I have followed the logician's lead, ascending yet again, to metamathematics, where both the symbols and their shapes are surveyed, the eye taking in everything below.

Gödel's theorem commences with the discovery of a scheme of coordination by which ordinary arithmetic may be given the power to talk about itself, the scheme reminiscent of Descartes's version of analytic geometry in which points in the plane are associated to pairs of numbers. The primitive vocabulary of formal arithmetic includes the primitive vocabulary of the predicate calculus but goes somewhat farther. Herewith a complete list of symbols organized into clusters.

Logical symbols:	~, ∀, ⊃, ∨, &, (,), S, 0, =, ·, +,
Propositional symbols:	*P, Q, R, S,*...
Individual variables:	*x, y, z,*...
Predicate symbols:	E, F, G, H,...

These symbols all belong to formal arithmetic, and they all carry latent fingerprints indicating that they simply have no meaning, but I have left off underlining, to spare the typesetter trouble. When I mean to draw attention to the very symbols themselves, back goes the underline.

Numbers are now assigned to the symbols of this system. The logical symbols get the numbers from one to twelve.

Logical symbols

~,	∀,	⊃,	∨,	&,	(,	),	S,	0,	=,	·,	+
↑	↑	↑	↑	↑	↑	↑	↑	↑	↑	↑	↑
1,	2,	3,	4,	5,	6,	7,	8,	9,	10,	11,	12

The propositional symbols are assigned numbers greater than ten but divisible by three.

Propositional symbols	*P,*	*Q,*	*R,*	*S,*...
	↑	↑	↑	↑
	12,	15,	18,	21,...

Individual variables? A number greater than ten, leaving a remainder of one when divided by three.

$$\begin{array}{llll} \text{Individual variables:} & v, & x, & y, \ldots \\ & \uparrow & \uparrow & \uparrow \\ & 13, & 16, & 19, \ldots \end{array}$$

Predicate symbols, finally, are assigned a number greater than ten which leaves a remainder of two when divided by three:

$$\begin{array}{llll} \text{Predicate symbols:} & \text{E}, & \text{F}, & \text{G}, \ldots \\ & \uparrow & \uparrow & \uparrow \\ & 14, & 17, & 29, \ldots \end{array}$$

Each symbol of the primitive vocabulary has now been assigned a tag, or Gödel number, as it has come to be called.

Under the aspect of formal arithmetic, symbols come together in specified and determinate ways. They may form formulas. Or they may combine to form sequences of formulas, one formula following another as in a proof. The numbering system not only assigns tags to symbols, it assigns tags to formulas and sequences of formulas as well. The formula $P \supset P$ has served well in the past as an illustration. Here it is pressed into service again. The numbers corresponding to symbols in this formula are 12, 3, 12. The formula as a whole is now given the number $2^{12}\ 3^{3}\ 5^{12}$, where 2, 3, and 5 are the first three prime numbers. This number is now assigned to $P \supset P$ as a whole.

Whole sequences of formulas figure in formal proof, as in the example from chapter five:

1. $\forall x(x + S(0) = S(x + 0)$	m_1
2. $\forall x(x + 0 = x)$	m_2
3. $\forall x(x + S(0) = S(x))$	m_3
4. $\forall x(x + 1 = S(x))$	m_4

The number to the right of each formula is its Gödel number. The sequence of four formulas is now assigned the number

$2^{m_1}3^{m_2}5^{m_3}7^{m_4}$. This is again a very large but perfectly determinate number, one capturing information in a powerful and compact arithmetical form.

There are two virtues to this numbering scheme. It determines that every symbol, formula, and sequence of formulas has a unique Gödel number, and it determines that every Gödel number uniquely determines some sequence, formula, or symbol of the system. The intermediary between what the system says, what it does, and what it can do is the fundamental theorem of arithmetic, which establishes that every number may uniquely be expressed as a product of prime numbers. Given the Gödel number assigned to $P \supset P$, on decomposition, $2^{12}\ 3^3\ 5^{12}$ and only $2^{12}\ 3^3\ 5^{12}$ emerges. And from $2^{12}\ 3^3\ 5^{12}$ and the original list, the logician can determine what formula the number represents. It is the formula $P \supset P$, as advertised.

What is unexpected in all this is that a way has been found for statements of arithmetic to comment about themselves. This is not simply a matter of cleverness, although the idea for the scheme is clever enough. The process of assigning tags to symbols endows arithmetic with a second voice. The first speaks directly to the numbers and their properties; the second, to facts *about* the numbers and their properties, so that uniquely among disciplines, elementary arithmetic stands revealed as a polyphonic discipline, its terse symbols suddenly acquiring a distant but distinct tonality.

The metamathematician now takes control of things. The formal sequence of formulas,

1. $\forall x(x + S(0) = S(x + 0))$	m_1
2. $\forall x(x + 0 = x)$	m_2
3. $\forall x(x + S(0) = S(x))$	m_3
4. $\forall x(x + 1 = S(x))$	m_4

he remarks, is a proof of the formula $\forall x\ (x + 1 = S(x))$.

What he has to say, he might say to the same effect by using Gödel numbers instead of words:

$$2^{m_1}3^{m_2}5^{m_3}7^{m_4} \text{ is a proof of } m_4.$$

But so far, of course, what he has to say remains within metamathematics. The words "is a proof of" are not words of arithmetic, but words *about* arithmetic. Nonetheless, discourse about arithmetic may be represented within arithmetic itself. After all, the words "is a proof of" express an arithmetical relationship between two numbers, one holding just in case the first number corresponds to a proof of the formula named by the second number. This arithmetical relationship may be reflected *completely* from within ordinary arithmetic by an arithmetical predicate—call it PR—such that:

$$2^{m_1}3^{m_2}5^{m_3}7^{m_4} \text{ PR } m_4,$$

or returning to classical notation,

$$\mathrm{PR}(2^{m_1}3^{m_2}5^{m_3}7^{m_4}, m_4).$$

This is a sentence indistinguishable in form from

Greater (7, 5),

which says that seven is greater than five. Both sentences are sentences *of* arithmetic. They talk about numbers, but by means of the code, the first sentence talks about other concerns as well—proof, most notably.

A third step remains, one involving double vision, as

$$\mathrm{PR}(2^{m_1}3^{m_2}5^{m_3}7^{m_4}, m_4)$$

is evacuated of meaning so that it becomes simply a line within the endless lines of the formal arithmetic itself, a series of shapes now, and not a statement of arithmetic at all:

$$\underline{\mathrm{PR}(2^{m_1}3^{m_2}5^{m_3}7^{m_4}, m_4)}.$$

The machinery for a very precise form of self-reference is now at hand.

RECURSION

It often happens that important ideas may be found loitering in the history of ideas without anyone quite realizing that the ideas are important, indeed, crucial. Recursion is an example. In its informal and habitual guise, recursion designates a step-by-step process of a sort already familiar through examples, various definitions specifying the grammatical formulas by moving up the inferential staircase. It is in the context of his great argument that Gödel came to recognize recursion for what it is, an expression in mathematical terms of the very essence of an algorithm.

Predicates (x is blonde) and relations (x is blonder than y) have already made a formal entrance in a formal language. A third class of mathematical entities is now required, and these are the functions. Functions are the living nerves of mathematics, taking objects to other objects, forming associations and binding elements of one set to elements of another. The mathematician endeavoring to make herself clear about their nature finds herself moving along a spiral of definitions, the impression she derives of resolutely reaching bedrock almost always mocked by a gray and uncompromising sense of uncertainty. Like the rest of us, attempting to explain the most ordinary of concepts, the mathematician is best served by example. Our domain is the world of the natural numbers. Every natural number may be multiplied by itself, 2 passing thus to 4; 4 to 16; 3 to 9; 5 to 25; and 50 to 2,500. Something is given and something is done with the result that something is found. What is given is a number; what is done is multiplication; and what is found is the number's square. The triple movement of the mind as it takes a number and squares it is represented mathematically by the action of symbols on themselves. Thus

$$f(2) = 4$$
$$f(4) = 16$$
$$f(5) = 25$$

all serve to convey the motion from a number to its square that has already been described in English and is now redescribed symbolically, with f designating the function itself, the leaning symbol applying pressure to a number in order to produce another number.

Nor need the numbers themselves be specified. With a variable in place of a specific numeral, the result is

$$f(x) = x^2,$$

a general all-purpose structure of command in which the mathematician records innumerably many mental operations, the open texture of the expression recording her decision to take *any* number and square it.

The arithmetical functions, as their name might suggest, are involved in the work of arithmetic itself. Addition is an arithmetic function, one taking *two* numbers to their sum; ditto multiplication; ditto division; and ditto all the rest.

In forging his proof, Gödel introduced a new class of functions into the mathematical empyrean—those that are *primitive recursive*. Other logicians had noted their existence, but Gödel ratified their importance. The primitive recursive functions are a special class of the functions in general; they are of extraordinary interest insofar as they are *mechanical* in a sense of mechanical that Gödel for the first time made precise.

Here is an example.

$$f(0) = 1$$
$$f(1) = 1$$
$$f(x + 2) = f(x + 1) + f(x)$$

Each number in the sequence generated by this function can be derived from the one behind it, like elephants clinging to one another's tail. The sequence begins with the numbers

1, 1,

this by means of the first two functions. The function that follows allows us to determine the next value in the sequence $f(2)$. But $f(2)$ is really only $f(0 + 2)$, which is $f(1) + f(0)$. These values we already know. The value of the function at 2 is 2. And at 3? There is no point in asking. Everyone understands.

Gödel's definition simply places this obvious example in the more general context of all functions that somehow map numbers onto numbers. The genius of the idea lies with its method. Instead of talking about an infinitely large object, recursion allows the mathematician to talk about a finite rule of construction. We are offered a recursive definition of a numerical sequence if in the first place the mathematician can specify the first number of the sequence and in the second place he can provide a rule defining the $(k + 1)$th number in terms of the kth number. Such is recursion's familiar loop.

The idea of recursion may be given a more formal definition. The primitive recursive functions begin with three simple functions.

1. *The zero function, Z:* Whatever number Z is given, the zero function returns 0: $Z(0) = 0$, $Z(1) = 0$, and $Z(10{,}000) = 0$ as well.
2. *The successor function, S:* Whatever number S is given, the successor function returns the successor

of that number: $S(0) = 1$, $S(1) = 2$, and $S(10{,}000) = 10{,}001$.

3. *The identity function, I:* Whatever number I is given, the identity function returns precisely the same number: $I(0) = 0$, $I(1) = 1$, and $I(10{,}000) = 10{,}000$.

These functions comprise *recursion's core.*
A definition now follows.

> *The primitive recursive functions are precisely those arithmetical functions that may be derived from recursion's core by a finite number of specified mechanical operations.**

Take, for example, the function $f(x) = x + 2$. Whatever the number x, this function adds to it the number 2, so that $f(4) = 6$, and $f(28) = 30$. Is the function primitive recursive?

It is. It may be derived from the successor function: $f(x) = S(S(x))$. The function $f(x) = x + 2$ is just the function that returns the successor to the successor of the number it is given.

Something has been given and something defined. Here for the first time an idea that has slithered secretly through history, slithering in fact from the Greeks through the minds of Leibniz and Peano, then on into the twentieth century, breaks into light for the first time. Recursion's loop has been given a formal definition.

THE CARDINAL CONTEMPLATES RECURSION

■ ■ ■ ■ Once while living in Vienna, I received a letter from the cardinal's secretary. It was written on the beautiful embossed stationery that the Church uses for its official correspondence

*See the appendix for details.

and addressed to Herr Dr. Professor. Knowing, the secretary wrote in elegant German, of my profound grasp of mathematical logic, would I, he inquired, be willing to explain the idea of recursion to the cardinal? From time to time, members of the institute where I worked were invited to speak to the cardinal privately; but generally the cardinal preferred to be instructed by the eminences of the scientific community and I could explain the secretary's letter only as the result of a bizarre miscalculation of my name.

I accepted nonetheless. Who wouldn't?

I met the cardinal's secretary at the antechamber of the cardinal's office in the Papal legation, which was located on a charming tree-lined street just off the Kärntnerstrasse. The secretary was dressed in the robes of a Jesuit; he had a small almost perfectly symmetrical face, the features drawn in dark colors: thick, but perfectly straight, eyebrows over black eyes; a narrow nose descending into dramatically flaring volutes; and a flat, thin mouth. His beard was so heavy that despite the fact that he had obviously shaved only recently, a bluish cast surrounded his cheeks and his chin.

The antechamber was baroque even by standards of the baroque in a very baroque city. There were pink and mauve celestial figures disporting themselves on the semicircular ceiling, which opened to a skylight, and pale intricate wallpaper on the walls, and an elegant red Empire couch facing a beautiful Empire desk. The creamy faded rug depicted a sad-eyed unicorn within a gate; it looked old enough to be authentic. A portrait of Pope John XXIII hung on one wall and on the other wall there was a magnificent late Giotto—the one in all the art books. The room was lit by the skylight and by a series of red-shaded lamps that were set on parquet-covered tables.

The secretary sat at the Empire desk and outlined the protocol of the interview. "His Eminence," he said, "wishes to understand the essentials. He is, of course, easily distracted by details. He will wish you to present the ideas clearly and concisely. He may ask questions of you, but you are not permitted to ask questions of him."

"Why, may I ask, does he wish to comprehend recursion?"

The secretary shrugged his shoulders beneath his very well tailored cassock. "His Eminence," he said, "is a man of wide curiosity."

The elegant standing clock began to chime the hour. It chimed three times and then trilled a few notes.

The secretary rose and ushered me into the cardinal's private office, holding the carved wooden door open before me and then immediately withdrew, closing the door behind him.

If the antechamber was splendid and baroque, the cardinal's private office was a study in sober discretion. Every wall was covered with old-fashioned built-in bookcases, the books almost all of them bound in leather or vellum, with titles, I could see, in almost every European language.

The cardinal was sitting at his desk.

Rising from his seat like a majestic red-robed seal, he leaned over to grasp my hand in his two hands and with a broad smile said that it was a great pleasure to see me again. I was flabbergasted. He motioned me to the high-backed wooden chair in front of his desk.

"It is kind of you to flatter me with your attention," he said suavely, his thick fingers folded over his rosary. "I am now past the age when learning is easily acquired, and although I have the curiosity, I lack, alas, the discipline to pursue study without the intervention of an interlocutor."

I allowed my eyes to roam the room, the books seeming to suggest anything but a lack of discipline.

"I would hardly think so, Eminence," I murmured.

"It is true nonetheless," the cardinal said, his voice now hardening just somewhat. "In this city where Gödel was born, it is scandalous that I should know so little of his achievements."

"Very few people do, Eminence, even today. It is easy to miss the moment when the wheel of intellectual history turns."

"Indeed," said the cardinal. And then he appeared to wait.

I had memorized a speech. "Gödel's theorem showed both that arithmetic is incomplete and that a proof of its consistency is beyond the powers of arithmetic itself. Some mathematicians—John von Neumann, for example—understood the proof at once and grasped its implications, but Gödel's reasoning was so subtle, and his proof such a masterpiece of concision and paradox, that at least thirty years were to pass before the general mathematical community understood that something remarkable had been achieved, some profound reorganization of reality undertaken."

The cardinal smiled broadly, moving his heavy cheeks as if he were hoisting them upward. "That is very nicely put," he said. "I was not aware that reality could be reorganized."

"I meant our perception of reality, Eminence."

"Between what we mean and what we say, there is always that fateful difference. Do you not find that that is so?"

"Of course, Eminence," I said.

"But now you must really tell me about recursion. Von Neumann himself explained the theorem's conclusions to my predecessor in this office. I remember the occasion well. You are, of course, too young to have known von Neumann. He spoke eight languages, all of them with an atrocious Hungarian accent. He was nonetheless lucid and composed. Perhaps you are not aware that he faced death with great terror. In the end, he accepted the sacraments. But he failed to explain recursion in his lecture, and reading this"—the cardinal pointed to a copy of *Gödel, Escher, Bach* that lay on his desk—"I find myself uncertain about its importance."

"Recursion," I said, "is a human way to grasp an infinite totality."

The cardinal looked at me steadily. "One requires an infinite mind to grasp an infinite totality. Is this not true?

"Gödel himself wondered," I said carefully, "whether the *human* mind might not be infinite."

The cardinal smiled again. "If that is so, why did he find it necessary to find a *human* way to grasp an infinite totality?"

It was a good question; it was also unanswerable.

The cardinal said, "Perhaps we may leave the grasp of infinity to the Holy Spirit. It is the human way that is of interest to me. I am, after all, only a man like other men."

"The natural numbers," I said, "go on and on."

"So it would seem."

"Recursion is a way of allowing *us* to go on and on as well."

"Perhaps an example?"

"Of course, Eminence." I began my second memorized speech. "The Fibonacci sequence is one of those mysterious consortiums of numbers that appear in mathematics and, once seen, appear in nature as well. The winding patterns of seashells, for example, obey a Fibonacci sequence; so do many other things. The sequence begins with the following numbers

$$1, 1, 2, 3, 5, 8, 13, \ldots,$$

and it is obvious that the numbers in the sequence are generated by adding the two previous numbers: 1 is the first number; 1, the second because $1 + 0 = 1$; 2, the third, because $1 + 1 = 2$; 3, the fourth, because $2 + 1 = 3$, and so on."

"I take it," said the cardinal, "that it is the 'and so on' that gives pause. And so on *how*?"

"There are two ways in which to give content to 'and so on.' One is by means of recursion." I placed on the cardinal's desk the pad of white paper I had brought with me and wrote the following symbols:

$$f(0) = 1$$
$$f(1) = 1$$
$$f(x + 2) = f(x + 1) + f(x).$$

"These symbols function as a recipe or guide. They offer the human mind a way of constructing the Fibonacci sequence one step at a time, beginning with 0 and going upward. They allow one to form a tower, one built from the bottom up."

The cardinal nodded his heavy head.

"The second method is to define the Fibonacci sequence explicitly, by means of an algebraic definition." I wrote down the standard formula:

$$x_n = \frac{1}{\sqrt{5}}\left[\left(\frac{1+\sqrt{5}}{2}\right)^n - \left(\frac{1-\sqrt{5}}{2}\right)^n\right]$$

The cardinal took the notepad from me and studied both formulas, his eyebrows tensing as he read each symbol.

"Both equations give the same answer, of course. At $x = 1$, $f(x)$ is 1, just as in the recursive equation. It is just a matter of a few algebraic manipulations."

The cardinal held up his hand. "I have confidence," he said. "Suppose that x were a very large number—a million, say?"

"The algebraic equation would give you the answer at once," I said, "but the computation would be tedious."

"And the recursive equation?"

"It wouldn't give you the answer at all, unless you already knew what $x = 999{,}000$ and $x = 998{,}000$ happened to be."

"Ah," said the cardinal, "then of what use is the recursive equation?"

"Suppose there were no algebraic equation at all or that the mechanics of working out the equation were too difficult. There is always the process of recursion, which allows you to reach a very large number by means of simple steps, repeated over and over again."

"Much like prayer," said the cardinal to himself.

"But there is something else," I said diffidently.

"You must tell me what it is."

"Recursion is an example of a *mechanical* process. The steps are broken down so simply that no thought is involved in carrying them out. One just adds the two previous numbers, the sequence building itself by itself."

"And the importance of this?"

"It gives precise meaning to the intuitive idea of *effective calculability*."

"Meaning?"

"Something that can be done in small discrete finite steps, which move toward a specific conclusion."

"In other words, mechanical?"

"In other words."

"We would appear to be moving in a circle."

"So it would appear, Eminence, but *mechanical* and *effective* and *calculable* are terms of ordinary language and as a consequence are vague or at least unclear. The recursive function I have defined is crystal clear."

"But its significance is clear only to someone who already knows the antecedent significance of *mechanical, effective,* and *calculable*?"

I began to shrug my shoulders and recognized the gesture as an impertinence. Somehow I managed to keep them from falling.

"One must begin somewhere."

"In some mystery or other?"

"Yes."

"It was clear to me," the cardinal finally said, "that recursion would be a subject to my taste."

At that moment, I could hear the clock in the cardinal's antechamber strike the hour and then tinkle its little song. The door behind me swung open and the cardinal's secretary stood in the space between the rooms, waiting respectfully. The cardinal rose again from his seat, extending his heavy hand. The interview was over.

■ ■ ■ ■

THE TAIL OF PROOF

What of incompleteness in all this? It has, so far, gone unmentioned, although I have spent quite a bit of time warming myself up. It is here, where this very question interrupts the reveries of my recapitulation, that the primitive recursive functions make their appearance on the historical stage of mathematics. Over the course of forty-five definitions, Gödel established that the essential arithmetic operations are primitive recursive and that almost all the metamathematical operations about the formal system are primitive recursive as well. *Almost all.* Not all. When I speak of the metamathematical operations about the formal system being recursive, I mean, of course, that they become recursive when they are coded by Gödel numbers and then expressed within arithmetic.

The result is a preliminary theorem. A recursive proposition about the numbers is one in which a recursive function or predicate figures. There is nothing mysterious in this. The proposition that 2 + 2 = 4 involves the recursive predicate of addition, and addition is recursive because within the confines of arithmetic, addition receives an interpretation along a familiar inferential staircase. That theorem now:

> *Every recursive proposition about the natural numbers may be expressed within formal arithmetic by a formula, and that formula is provable if and only if the proposition is true, and disprovable if and only if it is false.*

The simple statement that 5 is a prime number says something about the numbers, namely that 5 is prime. It happens to be true. It follows that there is a *provable* formula within formal arithmetic expressing that truth. And ditto for the ordinary arithmetical concepts, and ditto again for the ordinary metamathematical concepts, such as proof itself.

With this theorem, Gödel has tied recursiveness directly to the concept of proof, establishing, for all practical purposes, that arithmetic, *conceived purely as a recursive system,* is entirely complete, just as Hilbert had hoped.

Now for that undecidable proposition. Gödel proposed to construct a sentence of formal arithmetic that when given its ordinary meaning says of itself that it is not provable. This is an undertaking of three steps. The logician's voice must first be buried in arithmetic by means of the code, arithmetic must next be stripped of meaning by means of doublethink, and then doublethink must be reversed, so that a sentence of formal arithmetic comes to say of itself, Look here, there is no proving *me*. This is an astonishing polyphonic performance.

The details are as follows.

Let us consider formulas A(v) of formal arithmetic in which just one free variable v figures. The formula or expression "v is a prime number" is an example, one true if v is given the interpretation of 3 or any other prime number, and false otherwise. But v is free in A(v), and so we do not know the depth of its generality or whether it is intended to hold for all numbers, or for some, or for none at all.

These formulas, with their vacant free variable, may be arranged in a list, just as in Richard's paradox. There is the first such formula, and the second, and the third, and so on to the 223rd formula and points beyond. These are formulas of formal arithmetic. They do not yet convey meaning. They are shapes of the system and nothing more.

Every formula on the list is assigned its own Gödel number, so that looking at that list from above, the logician sees

formula (1) ← Gödel number for formula 1
formula (2) ← Gödel number for formula 2
formula (3) ← Gödel number for formula 3
.
.
.

formula (223) ← Gödel number for formula 223,

the list simply stretching on and on, or down and down.

If formula 13 is Fv then its Gödel number is $2^{17}3^{13}$.

A brand-new predicate A(v, x) now makes an appearance on the field of argument, one that holds of two numbers, v and x. Numbers, note. This is a predicate of ordinary arithmetic, a perfectly ordinary roly-poly relationship.

The predicate holds of any two numbers under two conditions. The variable v must stand for the Gödel number of some formula A(v) on the list, a formula in which v is itself free. This is an open-ended relationship. The logician has no idea yet what number v happens to be. And x must be the Gödel number of a proof of that formula, when the free variable v is bumped from that formula and replaced with a numeral, one expressing the number denoted by v, *whatever that happens to be.*

Suppose, for example, that the 93rd formula on the list is Ev—nothing more. Its Gödel number is just $2^{14}3^{13}$. The logician now invests this formula with its usual meaning, discovering that Ev expresses the arithmetical proposition that v is an even number. Not knowing what v is, the formula is indeterminate.

The predicate A(v, x) is brought into play. The variable v is the Gödel number of some formula on the list in which v is free. In this case, $v = 2^{14}3^{13}$. And the variable x is the Gödel number of a proof of that formula, when v is removed from the formula and replaced by $2^{14}3^{13}$. In that case, A(v, x) says that x is the Gödel number of a proof of the formula that results when the numerals expressing the number $2^{14}3^{13}$ are substituted for v. (The relationship A(v, x) is, of course, hardly the stuff of intuition.)

Now as it happens, the predicate A(v, x) is primitive recursive. And so it may be represented within a formal system purely by formal symbols—$\underline{\text{A}(v,\ x)}$, which peeps out from within formal arithmetic to say, when given meaning, what A(v, x) says from beyond formal arithmetic.

The logician next applies quantification to this perfectly ordinary formula of formal arithmetic, obtaining $\underline{\forall x{\sim}\mathrm{A}(v, x)}$. This formula, when properly interpreted (the logician breathing life into the symbols), says just what it seems to say. Nothing is a proof of the formula whose Gödel number is v when the free variable in that formula is replaced by a numeral naming the Gödel number of that formula.

But now for the infinitely cunning part of the proof. The formula $\underline{\forall x{\sim}\mathrm{A}(v, x)}$ has just one free variable. It follows that it, too, must appear on the master list of formulas with just one free variable. And so it does, in the 933rd position, say, just to make its appearance specific. Since it pops up on the master list, it must have its own Gödel number. And so it does. Say that its Gödel number is P, and say as well that $\underline{\mathrm{P}}$ is the symbol in formal arithmetic naming P.

And now, following the recipe that has governed the construction of the predicate A, let us substitute the numeral naming P for v in $\underline{\forall x{\sim}\mathrm{A}(v, x)}$, yielding $\underline{\forall x{\sim}\mathrm{A}(\mathrm{P}, x)}$. Or what comes to the same thing, let $\mathrm{A}(\underline{\mathrm{P}})$ stand in for $\mathrm{A}(v)$ on the master list. [$\mathrm{A}(v)$ is the same formula as $\underline{\forall x{\sim}\mathrm{A}(v, x)}$ and $\mathrm{A}(\underline{\mathrm{P}})$ the same formula as $\underline{\forall x{\sim}\mathrm{A}(\mathrm{P}, x)}$.]

The logician must now make these formulas speak, whistle, shout, and otherwise carry on. Just what does $\underline{\forall x{\sim}\mathrm{A}(\mathrm{P}, x)}$ say, when the logician invests its formal symbols with meaning? It says that there is no proof of a certain formula $\mathrm{A}(v)$ that results when the Gödel number of $\mathrm{A}(v)$ is substituted for v. Very well. Which formula? And with this question, Gödel's theorem and Richard's paradox meet, cordially shake hands, and go their separate ways.

The formula $\underline{\forall x{\sim}\mathrm{A}(\mathrm{P}, x)}$ says that nothing is a proof of the formula that results when $\underline{\mathrm{P}}$ is substituted for v in the formula $\underline{\forall x{\sim}\mathrm{A}(v, x)}$.

But $\underline{\forall x{\sim}\mathrm{A}(\mathrm{P}, x)}$ *itself* is precisely the formula that results when $\underline{\mathrm{P}}$ is substituted for v in $\underline{\forall x{\sim}\mathrm{A}(v, x)}$. And so $\underline{\forall x{\sim}\mathrm{A}(\mathrm{P}, x)}$, when interpreted, *says of itself* that it is not provable.

It follows that if formal arithmetic is consistent, there is at least one sentence within formal arithmetic that cannot be demonstrated. And what is worse, that sentence is true. Consider what it says.

There is in this the very strongest hint of paradox, but in the end, no paradox at all. Truth and proof diverge. There is no reason, save for wishfulness, that all of the truths of arithmetic should be provable within arithmetic.

The Hilbert program is doomed. It is now dead. The most elementary parts of our intellectual experience are incomplete.

This is, quite by itself, a stunning and surprising fact, one that constitutes a sharp rebuke to intellectual optimism. In the curious way in which surprises seem to generate surprises, a still more stunning fact is waiting in the wings. Some weeks after Gödel concluded his incompleteness proof, he noticed that his conclusions endangered consistency as well as completeness. The inferential trail now narrows itself, but it is easy to follow. The incompleteness theorem turns on the assumption of consistency, of course; within an *in*consistent system, anything goes and so nothing counts. Gödel's proof—the one I have just outlined—thus contains a strong central assumption: *If* arithmetic is consistent, *then* it is incomplete. The assumption comes packaged as a hypothetical and so involves two sentences. Let us say that consistency is expressed by the formula, **Formal Arithmetic is Consistent,** and then collapse these four words into one abbreviation—CONS. The second sentence collapses itself at once into a single formula—$\forall x{\sim}A(P, x)$, which says of itself that it is indemonstrable. The hypothetical that powers Gödel's argument thus has the overall form:

$$\text{CONS} \supset \forall x{\sim}A(P, x).$$

This is still a logical version of ordinary English, something found in the logician's metalanguage. But there is no reason that the same code that has served to express metamathematics within a formal system cannot be used to express this sentence within a formal system as well, whence

$$\underline{\text{CONS} \supset \forall x{\sim}\text{A}(\text{P}, x)}.$$

Inferences now commence. Is the consistency *of* arithmetic demonstrable *within* arithmetic? If so, $\underline{\text{CONS}}$ must be demonstrable. It says, after all, that arithmetic is consistent. But if $\underline{\text{CONS}}$ is demonstrable, so too $\underline{\forall x{\sim}\text{A}(\text{P}, x)}$, and this in a single step. But $\underline{\forall x{\sim}\text{A}(\text{P}, x)}$ is indemonstrable. Such is the burden of Gödel's *first* incompleteness theorem. It follows that $\underline{\text{CONS}}$ must be indemonstrable as well.

And such is the burden of Gödel's second incompleteness theorem. The consistency of arithmetic cannot be demonstrated from within arithmetic. Stronger methods are required, doubt displaced upward but not defeated.

Gödel discovered his second great theorem soon after he had concluded work on his first. He had discussed his conclusions with von Neumann and some weeks later von Neumann noticed the inferential trail from incompleteness to inconsistency that Gödel had already traversed. He wrote to Gödel at once to tell him what he had found, but Gödel had been there first and had already seen what von Neumann saw.

PRINCETON IN WINTER

A cold, mean, and wet wind blows through the streets of Princeton in late January. The sky is low and gray, but the ambient light is blue, and at twilight the northern sky is sometimes shot through with strange feral streaks of yellow. It rarely snows. This makes the winter seem raw somehow, like the edge of a wound. I walked these streets; I remember what they were like.

It was in winter that Kurt Gödel starved himself, the hospital records laconically recording the cause of death as "inanition." No one was surprised. The madness was an old friend come home. "I have lost," he said, "the power to make positive decisions. I can only make negative decisions." This is a remark that can only be understood by someone who has suffered depression. And so it can be understood by everyone. He had always been a valetudinarian, wearing heavy overcoats in the midst of New Jersey summers, fretting endlessly about his health. There are wonderful pictures of Gödel in the company of his great companion Einstein, sitting somewhere on a porch, perhaps at Mercer Street. Einstein is dressed in shirt-sleeves, his body rumpled and pudgy, his face ruddy in the still heat; Gödel sits next to Einstein, dry, precise, his hair slicked back from his forehead, his glasses hiding his eyes from the camera. And, of course, he is wearing an overcoat.

He believed that his enemies were trying to kill him by poisoning his food; he was aware of his delusion but could not control it. He drew the inevitable, the logical, conclusion, the paranoia forcing him to complete the fateful pattern that he could see enveloping him but that he could not evade.

He stopped eating. He entered the Princeton hospital in late December 1977. There had been talk of treatment in Pennsylvania; he refused the doctors. And why not? Einstein had died more than twenty years before; he had been an enormously reassuring figure, avuncular and appreciative. His friend and colleague Oskar Morgenstern has died earlier in the year. His mother was dead, buried on alien shores. His wife, to whom he was devoted, was ill. He had no wish to live alone. The last living links to the German language were gone. He had confessed many years ago that he could no longer follow, or follow completely, new results in mathematical logic. His flesh was wasting, the stark bones of his face giving him an owlish character. He spoke politely to Hao Wang by telephone in mid-January. Wang later remarked that he seemed to be speaking from a great distance, as if he had already crossed over.

He weighed very little now, perhaps sixty-five pounds. He was no longer hungry; he had never believed in time and now he had come to live beyond the reach of clocks. His skin grew thin, translucent. And then one day, sitting in his chair, as the sky lowered itself in the cold afternoon, he stretched out his hand and died.

APPENDIX: THE DETAILS

In order to make this definition entirely acceptable to connoisseurs such as ourselves, the logician need only specify those mechanical operations to which I made breezy reference in the text. *Aussitôt dit, aussitôt fait.* The first is composition, the second primitive recursion. The words "primitive recursion" are now functioning in a double sense, describing as an adjective the class of *primitive recursive* functions and designating as a noun the operation of *primitive recursion.*

With composition in charge, a function h is derived from two other functions f and g just in case $h(x) = f(g(x))$. The function $h(x) = x+2$ is derivable in this way from the function $f(g(x))$, where $g(x)$ is the successor of x, and $f(x)$ is the successor of the successor of x.

Primitive recursion is slightly more complicated. Thus far I have spoken only in terms of a single variable, the function $f(x) = x + 2$ taking one number and returning with another. The exchange undertaken by a function may also be coordinated with two variables and thus two numbers. The function $f(x, y) = z$ represents the form; and the function $Add(x, y) = x + y$, an example, one in which *Add* takes two numbers and returns with their sum.

Primitive recursion governs the generation of such functions according to the following scheme. A function h is derivable by primitive recursion from the function f just in case 1) $h(0) = c$, and 2) $h(S(n)) = f(n, h(n))$.

This is by no means as grim as it seems. The conditions in plain English: the value of h at 0 is fixed, the letter c designating a constant and so a particular number; and the value of h at the successor of some number n is determined by the double action of f acting on n and the value of h at n. It is easy to show that addition is a primitive recursive function using the operations of composition and primitive recursion. And easy to show that all of the ordinary arithmetical operations are primitive recursive as well.

The primitive recursive functions give precise expression to the idea of an algorithm. But there are some functions, such as the Ackermann exponential (don't ask), that are calculable, and obviously so, but not primitive recursive.

It is this circumstance that prompted logicians to the definition of the wider class of recursive functions. For this, an additional mechanical procedure is needed, that of minimization. A function $g(x, y)$ is given. A function f is derivable from g by minimization just in case $f(x) = y$, where y is the smallest number such that $g(x, y) = 0$. If there is no such y at hand, then f is undefined and lapses into irrelevance.

Thus let $g(x, y)$ be the function $Add(x, y)$.

$Add(x, y) = 0$ when both x and $y = 0$, since plainly $0 + 0 = 0$, and not otherwise. 0 is thus the smallest y enforcing a return to 0 by $Add(x, y)$.

It follows as a matter of definition that the function f is derivable from $Add(x, y)$ by minimization only if $f(x) = 0$. By the same token, $f(x) = 0$ only if $x = 0$ as well. At any other number z, $f(z)$ means nothing whatsoever.

The definitions of composition, primitive recursion, and minimization that I have given need to be stated in more general terms before they do all of the work that they are required to do. In particular, functions must be allowed to take any number of arguments, so that the definitions encompass $f(x_1, x_2 \ldots x_n)$ as well as $f(x)$. The details are available in any standard text—George C. Boolos and Richard Jeffrey's *Computability and Logic* (Cambridge University Press: 1974), for example.

But whatever the details, it is the *idea* of recursion that is important, the inferential staircase that starts somewhere and moves itself upward in finite stages; and even when the details have been mastered, or ignored, the idea remains, resolving itself like all great ideas into a vivid image of that staircase, those steps, and the infinitely weary attempt to climb them.

In everything that follows, when I refer to the recursive functions I mean the full class of recursive functions, and not simply the primitive recursive functions.

CHAPTER 7

The Dangerous Discipline

■ ■ ■ ■ The following story was told to me one evening by Irving Bashevis Singer in a coffee shop on Broadway and 85th Street; it was late at night and for some reason or other, I happened to mention Gödel's well-known remark that logic was a powerful discipline. "Listen," Singer said, "powerful maybe, dangerous for sure."

"Dangerous? How so?"

"I'll tell you a story," Singer said. This is what he told me. I have translated his words from the Yiddish:

There was a certain rabbi in the town of Yehupetz. He had a long, thin nose from which a drop was always trembling and a yellow beard that fell in tangles to his waist. Behind his back, the urchins of the village called him Rabbi Say Nothing, for his habit of hardly ever speaking. Young men about to be married would come to him for advice. The rabbi would sigh deeply, stare up at the ceiling of his study, lock his long fingers together, and mutter angrily without saying anything. When the elderly felt the wings of the Angel of Death beating about their bed, they would call for the rabbi. He would sit silently by their bedside, but beyond intoning the blessings, he would refuse to say anything more.

At home, his wife had long despaired of hearing the rabbi enter into normal conversation. "That one," she would say, tapping her forehead significantly, "has been struck dumb by the Evil One himself."

It had not always been thus. The rabbi had been famous for his ingenious interpretations of the Talmud.

He would read a section from the tractate *Bava Metzia,* which deals with the giving of gifts, and ask his students whether the right hand could give the left hand a gift.

"*Nu?*" he would ask, his eyes twinkling. His students sat there, concentrating on the question, but before any of them could answer, the rabbi would argue that if one could give to others, then surely one could give to oneself.

The students nodded.

Then in the next minute, the rabbi argued that this view was in fact completely mistaken. "*Nu*, blockheads," he said. "Can a man be taller than himself? Then how can he give himself a gift?"

His reasoning was so clever and his thin tenor voice so persuasive that his students were forced once again to nod their heads in agreement.

But then the rabbi would say, "On the other hand," and show that everything he had just said was wrong as well. Dancing from foot to foot in impatience, his earlocks flapping as he moved his head vigorously, the rabbi explained that even if one gives to oneself, time must go by between the giving and the receiving of a gift, and in that time a man becomes something other than what he was.

In truth, the Evil One, cursed be his name, corrupts each man according to a special plan. One day late in the afternoon, a peddler carrying a sack appeared in Yehupetz. He was a short, broad-shouldered man, somewhat stooped from carrying his heavy sack. He had no more hair on his head than a pumpkin, but he had enough hair growing from his nose and ears to cover two heads. He took his supper in the *kretchma,* eating a bowl of kasha and groats mixed with curdled milk, and afterward visited the bathhouse. His fingernails were long and yellow, and when he took off his shoes, the men of Yehupetz noticed that his toes were long and monkeyish. That night, he slept with the other schnorrers on a wooden bench in the study hall.

The next day he set off to peddle his wares. Unlike other peddlers, he did not stop and make small talk with the matrons who answered the door when he knocked.

"I have pots, two groschen," he would say brusquely.

"Your tongue, it wouldn't split in two if you said a few words more," the housewife would retort.

"If tongues had wings, peddlers would fly."

To the pious Jewish matrons of Yehupetz, who covered their heads with kerchiefs, the peddler sold tin pots, scullery knives, darning needles, thread, and tailoring wax, but when he knocked on a door and a Polish serving girl answered, he would plunge his monkeyish hands into his sack and pull out colored ribbons or embroidered kerchiefs or tortoiseshell combs.

The servant girl would look at these treasures and shake her head.

"But darling, there are many ways to pay me," the peddler would answer in perfect Polish.

His sack seemed to be bottomless. For the men of Yehupetz he had a device for removing ear wax, a brass ball shaped like the head of a lion that could be placed over walking sticks, warm gloves lined with rabbit fur, and a variety of liniments for chilblains, night sweats, and hemorrhoids; for the urchins of the town he had pennywhistles, candy, and cut beads, and for the young girls, silver bells, dolls with wooden eyes, and tiny mirrors made of polished brass.

And for the rabbi's Talmud students whose voices were just starting to croak, he had something special. He accosted them on the dusty street as they were going to the study hall in the afternoon. Holding up his hand, he said, "The Messiah has waited this long. A little longer he can wait."

The students stopped. The peddler reached into his sack and fished out a deck of playing cards. The oldest of the rabbi's students, a tall youth named Itche Bunzel, with a prominent Adam's apple and a long, thin nose like the rabbi's, took one look at the cards and spat into the street.

"My fine young gentleman," the peddler said, "these are special cards."

"The devil himself deals such cards in Gehenna," said Itche Bunzel.

"Can the devil's cards do this?" the peddler asked, and riffled the cards in his hand. As the cards jumped under his fingers, a plump woman wearing absolutely no clothes appeared on the card backs, as if by magic.

A sudden blush shot through Itche Bunzel's cheeks. He turned on his heel and stalked off, the other students following. But later that day, Gimpel the Goatherd observed Itche Bunzel sitting on a rock in the field above the village mill. He was trying to riffle through a deck of cards.

By the end of the week, the peddler was able, as the saying goes, to keep two kopecks company in his pocket. On Saturday morning he attended services and placed five groschen in the poor box. After fasting all day, he took his evening meal at the *kretchma,* where he ordered a carp head boiled with turnips, pickled beets, stewed onions, and pears in honey with almonds. The rabbi's students reported to him that the peddler no longer slept with the other schnorrers on the study-hall benches but had asked for a bed at the *kretchma*.

The rabbi asked Reb Avigtor to summon the peddler to his study.

"At once, Rabbi," said Reb Avigtor and rushed off.

The rabbi was seated at his long wooden study table smoking his pipe when Reb Avigtor knocked and opened the door slightly.

"We're here, Rabbi," he said.

"So I see."

The rabbi motioned the peddler to a chair. The peddler sat facing the rabbi, fingering his beard and poking at his ears. The rabbi coughed decorously.

As if he were reading the rabbi's mind, the peddler asked abruptly, "So where is it written one must be poor?"

The rabbi said, "Where is it written that cows must walk on four legs? Nevertheless, cows can't fly."

The man said nothing in response, but it seemed to the rabbi that there was something mocking in his thin smile. Then the peddler reached into his sack and withdrew a leather-bound book and placed it on the rabbi's desk.

The flame on the rabbi's candle sputtered in a sudden gust of wind. The rabbi felt a pain in his nose and then a pain in his ears. His hands and feet tingled. He felt the urge to cough; drawing out his handkerchief, he sneezed violently instead. The rabbi wondered if an imp had taken possession of his soul. And then, in an instant, the pains in his nose and ears vanished. It was revealed to him that everything was as it should be. A feeling of joy flooded his soul. The peddler continued to sit where he had been sitting.

The very next day, the peddler was seen departing Yehupetz, his heavy sack slung over his shoulder. A week went by and then another. When the rabbi returned to his students, he observed a new power of analysis in himself. Passages in the tractates that had long been mysterious were suddenly clear. He saw meanings behind meanings. The words would form and re-form themselves on the printed page. Arguments spouted from his mouth like teeth. It appeared that Jonah had swallowed the whale and that a secret message lay coded in the Song of Songs. Passages that his students had assumed were about the turning of the seasons or the laws for ritual slaughter, the rabbi showed had significance for the design of the universe or else they contained secret plans for the construction of Gehenna.

The students could barely follow the intricacies of his thoughts. He would pose the simplest of questions and then reveal complexities behind the plain meaning of the Hebrew words. "Why is it written that Angels have two wings? Why not four?"

The students would grope for an answer. The rabbi would interrupt them. "Blockheads," he cried, "do you not see that the Almighty created being out of nothingness with only two numbers?"

In time, word of the rabbi's powers spread far beyond the town of Yehupetz. Students came to listen to his discourse from as far away as Chelm and even Warsaw. He would no longer enter into disputation with students, but carried on both sides of the discussion by himself, saying, "On the one hand," and "On the other hand," as if he were two people instead of one. More and more, the rabbi came to study the Talmud in order to be able to formulate arguments, rather than formulating arguments in order to study the Talmud. He designated certain of his students for special study. Meeting with them in the study house after he had dismissed the others, he would demonstrate all manner of unusual propositions, showing how by means of logic words could be made to deny what they seemed to assert and assert what they seemed to deny. Before his intimates, he was more than brilliant. His eyes seemed to burn with a strange fire.

One night, he walked home through the dusty streets of Yehupetz. As was his custom, he took only bread and an apple with honey for his evening meal. Afterward, he mounted the narrow wooden steps to his study. Seating himself at his desk, he lit his pipe. A great cloud of smoke filled the room and tickled his nose.

Beyond the study window, fruit trees were in bloom. Birds were twittering in the orchard. Insects whirred and chirped in the warm evening air.

The rabbi peered at the familiar surroundings of his study: the bookcases with their leather-bound volumes, the silver candlesticks, the wooden candleholder on his desk, where his wife had lit a sputtering taper, the worn woven rug on the oak floor. He felt a pain in his nose and then again in his ears. There was a burning underneath his heart. The silver candlesticks that his father, the rabbi of Yehupetz before him, had bequeathed to him, stood upright as sentinels. It seemed to him that all the things in the world were merely symbols and that the universe was arranged as a gigantic argument. The pale stars in the sky beyond his study window were asserting, "On the one hand . . . ," and the crescent-shaped moon was replying, "But on the other . . ." The leather-bound books in his bookcases were chattering with one claim and the very smoke in the air was answering with another. Everywhere things were in dispute and everywhere there was an argument, every part of the firmament talking, observing, contradicting, defining, mocking, looking about in feigned astonishment, pointing things out.

When the rabbi awoke, the candle had burned itself out.

The next morning, the rabbi called his initiates together. "There is no truth," he said, "only argument."

His students regarded him uneasily. "Where is it written," the rabbi asked, "that there must be truth?"

The rabbi then proceeded to argue that there could be truth in the world only if it were true that there was truth in the world.

The students nodded.

"Blockheads," the rabbi said, "what good is truth if truth is needed to find it? Why should a man fish for carp if he needs a carp to fish?"

The students nodded again.

From that day forth, the rabbi devoted himself to arguing against every settled interpretation of the Talmud. There was nothing he could not do, no position he could not uphold. Whenever he reached a particularly puzzling interpretation, he showed that, contrary to expectation, it really was in accord with the text and with what Rabbi Akiba or the other commentators meant to say. He showed how everything followed from contradictions and how contradictions followed from everything. As he spoke, he paced back and forth in the study hall, his earlocks flapping, his gabardine coat open. He stabbed at the air with his forefinger and pulled on his earlobes. He spoke so rapidly that on several occasions he seemed in danger of swallowing his own tongue.

His arguments grew more and more outrageous. It was as if an imp or a demon possessed him. He argued that there was no obligation to keep holy the Sabbath and that God had never made a covenant with Moses. He argued that the dietary laws were instituted to be violated and that the Almighty meant man to eat pork and that meat and milk were meant to be taken together. He argued that women who were unclean were clean and that women who were clean were unclean.

His disciples and initiates grew troubled, but they were afraid of his ridicule and the power of his tongue. Reb Avigtor said to him, "There are things that are given to us to know and things that are given to us not to know." The rabbi turned and in a torrent of words argued that what was known was false and what was not known was true. Reb Avigtor blanched.

In truth, the rabbi lived only for the moment he could appear before his circle of initiates. He slept little and took only a few mouthfuls of kasha for nourishment. His beard grew tangled and his eyebrows bushy. Small insects appeared in the pockets of his gabardine coat. When he walked down the streets of Yehupetz, the village cats slunk away with a hiss. At night, a strange fire seemed to glow in his study and bats were seen to beat about the study's windowpanes.

Before his initiates, he now argued that the pious man was wicked and the wicked man was pious. He argued that if nothing was forbidden, everything was permitted, and then he demonstrated that in truth nothing was forbidden. He argued that God had made the world to mock men and that men had no other recourse than to mock God. He argued that pleasure was the highest good and that self-indulgence was a form of righteousness. He argued that the Messiah had come, and he argued that the Messiah would never come. He argued that the Torah had been written by men and that men could do anything that the Almighty could do. He argued...

And here, Irving Bashevis Singer stopped. I waited for him to continue.

"There is no more," he said. "A man has to know when to be silent."

■ ■ ■ ■

CHAPTER 8

Flight into Abstraction

Logic has always been a dangerous discipline, any number of logicians going mad after finding themselves hopelessly lost in the wilderness of their own thoughts. When years later they are rescued by the thumping helicopter of common sense, they pick gratefully at emergency rations, and smile for the cameras, but when asked what on earth they thought they were doing, they can do little more than shrug their bony shoulders, saying, if they say anything at all, that like Gödel himself they were looking for something they could not find. The great logicians contained the impulses toward madness, living lives of tense equilibrium. Like philosophy itself, logic remains now and forever a strange undertaking, the symbols of the discipline alien, even to mathematicians, and made stranger still by the service into which they are pressed—the delineation of the *laws of thought*. This last phrase should be imaginatively weighed against the physicist's *laws of nature,* the physicist's task, the brokerlike business of discovering how a semiconductor, a falling apple, or the universe works, while the logician investigates the laws by which the laws of nature work, pursuing himself by himself in the endless night.

It is against this background that Gödel's great theorem ignited with a sibilant hiss in 1931, the full force of its concussive power destined not to be felt fully for at least sixty years; and it is this air of intellectual menace and mystery that must be contrasted with the physical environment in which it found itself improbably revealed—Princeton in the 1930s. Gödel made his results known in Vienna, of course, and a handful of logicians understood at once the depth and nature of his results, von Neumann especially departing from his honeymoon suite to talk with Gödel (his bride no doubt wondering disconsolately at his odd commitment to Higher Things); but it was to Princeton's newly formed Institute for Advanced Study that Gödel came several times during the decade before settling into the institute permanently in 1940, and the circumstances of the great European intellectual diaspora that gathered force during the 1930s made it inevitable that the weight of his insights would be felt most profoundly in the United States.

I do not think that Princeton was then appreciably different from the place I knew thirty years later. Great elms everywhere, a sense of lavish leafiness; old, red, vaguely gothic buildings, some of them dating back to the eighteenth century; richly grassed broad green lawns; the blonde, blue-eyed, disturbingly adult-looking undergraduates wandering in and out of their handsome porticoed eating clubs, dressed in white flannels and jackets bought at J. Press; and beyond the campus itself, the placid plush of the somewhat moth-eaten, mosquito-infested countryside of rural New Jersey, marshy where the land is low, thickly wooded where it is not; a few dim shabby towns straggling along the bus route, only the bar and the single restaurant lit at night by Edward Hopper; Trenton to the south; and impossibly far ahead, the lights of life, New York, two hours distant.

The institute occupied ground somewhere north of the graduate college. Einstein was there, and so, too, John von Neumann and Oskar Morgenstern, at least on a temporary basis; the place was intended to catch the cream of European refugees; but whatever the ambience at the institute, it was still an institute at Princeton and still a drab collection of wooden structures in which the German-speaking atmosphere of *Mitteleuropa* merged improbably with a university town that a little over ten years earlier had seen F. Scott Fitzgerald come and go.

Gödel lectured on his own results, of course, and while the mathematicians (and philosophers) at Princeton for the most part could not and did not understand a word of what he said, a group of young American logicians—Alonzo Church, Stephen Kleene, J. B. Rosser, a few others—sat in his audience and took notes, Gödel standing there owl-eyed, writing out the details of his great theorem on the blackboard, answering questions about notation and the subtle details of his proof, while the very men who were to make mathematical logic into a majestic discipline sat in cramped wooden seats, too big in their padded thirties for chairs meant for slim-hipped undergraduates, staring at the blackboard, conscious, as they must have been, that whatever would happen to them later in life, there at that very moment, they could see the flame of genius flash and flicker furiously.

Gödel's incompleteness theorems were what they were: absolutely decisive. They at once endowed mathematical logic with complete mathematical maturity and emptied the Hilbert program of its conceptual interest. By means of talent and timing, Gödel had compressed the usual stage in which a new discipline develops into a matter of a few years, taking mathematical logic almost instantaneously from the threshold of baffled flutter and probe to an entirely new level of sophistication.

With the annunciation and communication of the incompleteness results, the axis of interest along which mathematical logicians had walked began to rotate in space. Gödel had shown that arithmetic was incomplete and he had shown moreover that the proof of its consistency lay beyond arithmetic itself. These were demonstrations as much a part of the intellectual structure of this century as general relativity.

But they were *demonstrations*; they showed what they had shown, and this beyond the shadow of doubt. Considering that the Hilbert program had been intended to ratify the mathematician's confidence in mathematics, it is ironic that the theorem that ended those expectations was itself a superb, an irrefragable, example of mathematical certainty; the theorem in doing what it did managing to say what it said in two different ways. Almost at once and almost imperceptibly, interest turned from what was finished and final to the unexplored country opened up by Gödel's proof.

ALONZO CHURCH AT PRINCETON

The movement of ideas owes something to the quirks of personality. Late in his life, Gödel remarked modestly to the logician Georg Kreisel that he was sure when he discovered his incompleteness theorems that it was but a matter of time before someone else would have discovered them. In this, he was almost right. The essential facts about the incompleteness of arithmetic follow from Alfred Tarski's work on the concept of truth in formalized languages. And this suggests a relentlessness in the accumulation of an idea, almost as if that idea were moving through history on its own. But what

Gödel neglected to mention was the overwhelmingly particular nature of his own proof, its rare combination of elegance and precision, and the stunning nature of the coordinate method he discovered. No one but Gödel could have injected that magnificent object into the stream of time, and no one has ever suggested otherwise.

From a general point of view, Gödel's theorem demonstrates that some things are impossible. The mathematician cannot write axioms governing the whole of elementary arithmetic, and he cannot do it because it cannot be done. Logic has revealed an edge of the universe. Other theorems were to follow in the 1930s. Taken collectively, they represent an extraordinary contribution to the history of human self-analysis, showing, if nothing else, that while logic could not meet the request for certainty, it was the request itself that must have been defective, or at the very least, unsophisticated, the demonstration of its defects far richer and more compelling than the request itself. From the point of view of the idea commanding this book, Gödel's theorem did more: it gave a natural, persuasive account of a class of mathematical objects, one that for the first time pulled the concept of the algorithm into the community of common mathematical concerns. Logicians had talked of mechanical procedures, and they had exhibited mechanical procedures in their work, and they had in a loose and informal fashion appealed to the very idea of a mechanical procedure in constructing various logical programs, but in defining the class of primitive recursive functions Gödel, like all great logicians, excavated its logical core, and gave that core an enduring mathematical representation.

The work, and the definition that it yielded, reflects the powerful imprint of Gödel's own sensibilities. It is precise, down-to-earth, and commonsensical. For all of his unrivaled power as a logician, Gödel was by nature conservative, preferring to express his mathematical thoughts in a familiar language so long as they could be expressed in that language and venturing beyond only when compelled to do so by extraordinary circumstances.

It is all the more remarkable, then, that the second definition of a mechanical procedure should issue from an entirely different personality.

Alonzo Church was my teacher at Princeton, and although I realize he was then roughly the age that I am now, I still think of him, when I project myself backward to 1966, as if he had achieved a kind of fantastic temporal bulk, one that reflected not so much his actual age as the curious and incidental fact that the slow, deliberate gross motions of his body were under the control of what appeared to be a purely reptilian brain, the higher cerebral functions entirely given over to mathematical logic, and by means of this advantageous division of responsibilities, in which he was part troglodyte and part logician, he had managed to thicken the flow of time so that his years were other men's decades, his decades, their lifetimes.

He was in his work deliberate, abstracted, detached, and precise, embodying completely the purest ideal of Platonic pedantry in his papers, and in his magisterial textbook, which contained more than four hundred footnotes, some of them sprouting footnotes of their own, like secondary tumors, and he was in daily life, deliberate, abstracted, detached, and precise, as well, exhibiting in the usual oiled give-and-take of conversation a kind of explicitness that effaced the distinction between what he did and who he was, so that when asked, while standing at a window, whether it was raining, he would be certain to indicate that it was, *if* it was, but that it was raining *outside*, his baffled interlocutor, often a departmental secretary, seeing in this remote assiduity only signs of an ineradicable intellectual perversion.

His work habits were legendary. He would rise at twelve, or so it was said, and simply work throughout the day, the evening, and most of the night, sitting implacably at his desk, writing his manuscripts by hand and editing the papers submitted to the *Journal of Symbolic Logic* entirely on his own; and to this immense industriousness, he brought as well a kind of intellectual power that while it was not by any means a form of genius represented nonetheless a collocation of electrical currents, one that could be seen flickering somewhere beneath the great gray and gloomy ooze of his indefatigability.

He was married; he had children; he must have participated in, or at least heard, I suppose, the rich booming rhythms that make the world beyond mathematical logic shiver and shake, but while I can imagine Church displaying kindness in an offhand incidental way, stooping slowly to retrieve an errant paper for his secretary or accepting with grave unconcern the excuses of his graduate students, I cannot by means of any mental motions of my own displace the center of his passion from mathematical logic to any purely human form of concourse.

He was not eccentric; simply massive.

And yet, now that I look back upon his work, I can see far more clearly that Church's pedantry was as much a form of protective coloring as a natural wardrobe, for the man's mind was in some sense intoxicated by abstractions, dominated by a powerful urge reflected in everything he did to get beyond the debris of the commonplace and the detritus of things and make contact with a universe that is spare, clean, simple, elegant, and profound.

Just look at the calculus of lambda-conversion.

THE CALCULUS OF CONVERSION

The recursive functions that Gödel introduced in his incompleteness theorems give one voice to the concept of an algorithm; they embody some aspect of effective calculability, the notion, passing like signals in smoke from Leibniz to the twentieth century, that reasoning effectively, which is entirely a human achievement, can be demarcated and so defined, the reasoning stripped of its mystery by means of a series of mechanical steps. The recursive functions *are* mechanical: they *do* move in steps upward from zero; they could have been implemented by a machine in 1931, if only there were a machine to implement them. And they are very easy to grasp.

Alonzo Church introduced his calculus of lambda-conversion to the small community of logicians in 1936. It is one of those pieces of mathematics that at first seems pointlessly complex and pointlessly abstract, the enterprise suffering the curse of double desuetude. And yet curiously enough, I had better say at once, the calculus has both practical point and purpose; having been created by Church in order to articulate *his* vision of effective computability, the calculus turned out many years later to be instrumental in the development of various computer languages, another queer and troubling example of pure thought preceding its own instantiation in matter.

Like a lattice, the calculus of lambda-conversion intersects itself at indefinitely many places. It is difficult to grasp, difficult to understand, and difficult to manipulate. Still, the fundamental idea is relatively straightforward, especially when it is given in an informal context. Something is given—a data set, say, or a list, or a group of numbers. Call it A. And something acts on the data. Call it F, so that FA represents the result of something active, that F, acting on something passive, that A. The yin and yang of the calculus thus resolve themselves into an old human pattern. Such is what is called *application*, one of the two basic operations of the calculus.

There is, next, *abstraction*. Suppose that x is an ordinary variable, varying as ordinary variables do, designating now this and now that; and suppose, too, that $M[x]$ is a larger expression, one which in some way depends on the value of x, as in the example of

x smokes,

where the sense of the whole expression depends on just who x is.

In mathematical terms, the same relationship of confused identity and dependence is revealed when $M[x]$ recaptures its familiar form in an expression such as

x^2,

where the value and so the identity of M depends on the value and so the identity of x; and I say the relationship is confused because under ordinary circumstances, it is ambiguous. Is x^2 greater than 10? The question is pointless pending a resolution of the identity of x. If x is equal to 3, the answer is no, but if x is equal to 4, yes.

But what, on the other hand, of the question whether x^2 is always less than $x^2 + 1$? The variables retain their indeterminacy; not so the question itself, which now pops up unambiguously as a gimlet-eyed query, one admitting of a single obvious answer: x^2 *is* always less than $x^2 + 1$.

Something has gone slightly wrong, some perspective has been violated, as the mathematician's vernacular reveals an unsuspected difference between x^2 when considered as the value of a certain function, and x^2 considered as the function itself. We are talking, in the first case, of what x^2 denotes; but we are talking in the second case about the infinitely long relationship between numbers that x^2 specifies no matter what x happens to be. The difference is one between things and relationships. It is this second sense of x^2 that Church's calculus is intended to capture.

The lambda of lambda-conversion now makes a semiformal entrance in Church's scheme, the expression

$$\lambda x.M[x]^*$$

denoting the relationship between x and $M[x]$, this for varying values of x, so that

$$\lambda x.x^2$$

designates the relationship between every positive number and its square—the *relationship* itself, note, and not any of its values. If one thinks of the relationship between x and x^2 as comprising an infinite list of pairs (2, 4; 3, 9; 4, 16; and so on), then it is that list as a whole that $\lambda x.x^2$ represents. Lambda is an abstraction operator, reaching into the matrix of an ordinary mathematical expression and pulling to the foreground its abstract essence; like the universal or the existential quantifier, lambda *binds* the variables under its control.

In a purely mathematical context—ordinary mathematics now, nothing fancy—application and abstraction are combined to form an extraordinarily flexible system of notation. In

$$(\lambda x.x^3)3 = 3^3 = 27,$$

the symbols say that $(\lambda x.x^3)3$ denotes a certain map or relationship, the one that in *this* case sends 3 to its cube, and in the following case sends 4 to *its* cube

$$(\lambda x.x^3)4 = 4^3 = 64.$$

Here x^3 is playing the role of M, varying within itself as x varies.

Like so many other things that we take for granted, sym-

*In this and what follows, the dot serves as my own mark of punctuation; otherwise, parentheses would multiply pointlessly.

bols have a strange life of their own, sobbing and stumbling about the page just when we expect them to be circumspect, or lying there inertly when we most wish that for heaven's sake they would say something intelligible. So, too, the symbols of Church's lambda calculus. For one thing, they are *type free*. The term is the logician's own, and it designates a certain freedom within a notational system that obtains when the system's operations may be self-applied without fear of contradiction. Ordinary English affords ordinary examples of type-free operations, as when we say that John believes that snow is white and then say again that John believes that he, *John*, believes that snow is white, the very indication of belief appended to itself a second time, so that there are now two beliefs gamely ricocheting from one place.

But if ordinary English is type free in one respect, it is *not* type free in others, the various paradoxes showing that certain forms of iteration engender contradiction while others verge toward incoherence. In Church's symbolic world, there is appending without limit, so that when F has been applied to A it may again be applied to F, as in FFA.

It is this superb flexibility that makes the lambda calculus a subtle instrument of arithmetic definition. Application ties the calculus to specific objects, but whatever the tie, they are, those objects, functions and not numbers, so that

$$(\lambda f)[f(f(x))]$$

denotes the function f such that f operates on itself. Further applications may serve to tie this function to yet another function—g, say, where $g(x) = x + 1$. Then the expression $(\lambda f)[f(f(x))]g$ specifies the double action of g, so that

$$(\lambda f)[f(f(x))]g = g(g(x)) = g(x+1) = x + 2.$$

Ordinary English conveys the sense of these symbols in terms reminiscent of an old fashioned horror movie:

(The Thing) [whose nature it is to be iterated] applied to the Thing [whose nature it is to add 1 to any given number x] yields $x + 2$.

The pattern invoked in this purely verbal formulation admits of an even simpler overall description, with the operations of abstraction and application covered in one fell swoop by three majestic words:

(*Specification*)[**Identification**] *application,*

the three operations determining an abstract world, one in which functions signal to one another against an otherwise blank background.

This is apt to seem arid as well as abstract. Nonetheless, the triplet of specification, identification, and application is part of an immemorial human pattern as well:

Who's there?

Me.

Oh.

A FORMAL WORLD

The lambda calculus whose elements I have just surveyed is simply a logical instrument, a way of describing things; but Church envisioned the calculus as a formal system, and it is only when the struts are pulled away from the system that Church's blimp takes off and soars. In comparison with the predicate calculus, indeed, in comparison with all other formal systems, the lambda calculus is a masterpiece of spare precision.

There are only three symbols to begin with

λ, (,),

and of these, two are merely marks of punctuation, an unavoidable concession to clarity.

There is in addition an infinite list of variables,

$$a, b, c, \ldots, x, y, z, a', b', \ldots,$$

the various notational devices simply there in order to allow the logician to say what he would say without limit.

And there is absolutely nothing else.

Using one real symbol (λ), a handful of variables, and his own powerful imagination, Church proceeded to create a universe out of thin air.

The formation rules of the system are straightforward and they specify not only which sequences of primitive symbols are legitimate but also which variables are free and which are bound. The usual distinction between the logician's own language and the formal system he is describing is now in place. The boldface letters do not appear *in* the formal system. They have been invited to the party in order to talk *about* the system instead. The voice implementing the formation rules is Church's own:

1. A variable $\mathbf{x}$ is a well-formed formula and the occurrence of the variable in this formula is free.
2. If $\mathbf{F}$ and $\mathbf{A}$ are well-formed formulas, then $(\mathbf{FA})$ is well-formed, and an occurrence of a variable $\mathbf{y}$ in $\mathbf{F}$ is free or bound according as it is free or bound in $\mathbf{F}$. Ditto for a variable $\mathbf{y}$ in $\mathbf{A}$.
3. If $\mathbf{M}$ is well-formed and contains at least one free occurrence of $\mathbf{x}$, then $(\lambda\mathbf{xM})$ is well-formed and an occurrence of a variable $\mathbf{y}$, other than $\mathbf{x}$, in $(\lambda\mathbf{xM})$ is free or bound in $(\lambda\mathbf{xM})$ according as it is free or bound in $\mathbf{M}$. All occurrences of $\mathbf{x}$ in $(\lambda\mathbf{xM})$ are bound.

These rules proceed by recursion and so they proceed from the ground up, the logician, stout glandular Church, specifying one class of symbols absolutely—the variables standing alone are well-formed: period—and then specifying the rest of the well-formed formulas in terms of well-formed formulas already specified.

The rules are tedious, but they are precise. A variable standing alone is well-formed. Well and good. That means that

$$a$$

is a legitimate expression of the calculus. There it stands. Alone. It is free and so unbound, like a glass waiting patiently to be filled.

If the formulas ***F*** and ***A*** are well-formed—

Do we *know* that ***F*** and ***A*** are well-formed?

We do not. But *if* they are, then so is (***FA***). If not, not.

How do we determine *whether* ***F*** and ***A*** are well-formed?

By progressing downward.

F is well-formed *if* its parts are well-formed; if it has no parts, ***F*** is well-formed if ***F*** is a variable standing alone. So either ***F*** is composed of still further formulas G and ***H***, such that ***G*** and ***H*** are well-formed, or it is composed of a single variable, standing alone. If neither of these options holds, ***F*** is not well-formed. So then, neither is ***FA***.

Finally, specification by means of lambda results in a well-formed formula (λ***xM***) just in case ***M*** is itself well-formed. Recursion again and the old familiar staircase.

Freedom and bondage proceed apace. It is lambda (λ***x***) that *binds* the variable ***x*** in ***M***, reaching into the matrix of the formula to grab the variable and hold it fast. With (λ***x***) preceding ***M***, (λ***xM***) achieves a specificity of intention denied to ***M*** alone, (λ***xM***) taking charge of the operations at hand in order to say and so to stipulate that reference is being made to that function described by ***M***—none other.

Note what is remarkable in all this: no quantifiers, no predicates, not much of anything, a world as severe and sculpted as a marble torso.

CONVERSION

The symbolic sky is filled now with functions and so comprises a world in which everything is abstract and absolutely nothing is concrete, rather as if a court of law were to convene without ever hearing cases or seeing witnesses, the judges more than happy to allow their attention to be volleyed from citation to citation, passing without concern from Dinklesbury *vs.* The State of Connecticut, which deals with issues of indemnification in the law of torts, to Breaburn *vs.* Breaburn, a famous criminal case of the early eighteenth century, in which—but I digress. In the calculus of lambda-conversion, there is no mechanism yet for volleying anything anywhere. The universe is static. Nothing moves. And the functions? They are simply sitting there.

It is *conversion* as an operation applied to symbols that introduces forces into this world, formal objects acquiring the power to alter themselves into other formal objects, so that a universe of static relationships slowly undergoes a transformation, the abstractions changing in nature before the logician's eyes, almost as if she were watching a logical Lava lamp.

There are, in all, three conversion rules, and each is similar in scope and effect, governing as they do the act of substitution. Within the confines of that act, the logician is allowed

1. To replace any part M of a well-formed formula by substituting y for x in M—provided that x is not a free variable of M and that y does not occur in M.*

*Restrictions on the replacement of free and bound variables are here for the same reason they were present in the predicate calculus—gibberish control. As before, I have placed the restrictions on the rule in brackets in order not to deflect the reader's attention. It is worth mentioning, though, that the proper definition of substitution turned out to be an immensely difficult intellectual task, one that almost every great logician bungled at one time or other.

It is rule number 1 that permits the transmogrification of

$$\lambda x.x^2$$

into

$$\lambda y.y^2.$$

The rule allows the logician a certain crucial indeterminacy of specification, so that the logical truth that $\lambda x.x^2 = \lambda x.x^2$ does not hinge on the choice of x alone, but holds whatever the variable happens to be.

If rule 1 handles an obvious exchange of variables, rule 2 handles the larger project of exchanging larger items in the calculus. Thus the logician is allowed

> 2. To replace any part $((\lambda \boldsymbol{x} \boldsymbol{M})\boldsymbol{N})$ of a formula by substituting $\boldsymbol{N}$ for $\boldsymbol{x}$—provided that the bound variables of $\boldsymbol{M}$ are distinct both from $\boldsymbol{x}$ and the free variables of $\boldsymbol{N}$.

M and N, as they appear within the confines of $((\lambda M)N)$ represent formulas of whatever degree of complexity.

Peering inside $(\lambda \boldsymbol{M})$, for the moment, let us suppose it has the form $(\lambda \boldsymbol{x}.\boldsymbol{M}\boldsymbol{x})$. The variable $\boldsymbol{x}$ is bound in this formula. It lies within the confines of a lambda operator. However, $\boldsymbol{x}$ is free in $\boldsymbol{M}$ itself. No operator binds it. Suppose *this* formula is adjoined to $\boldsymbol{N}\boldsymbol{x}$, yielding $((\lambda \boldsymbol{x}\boldsymbol{M}\boldsymbol{x})\boldsymbol{N}\boldsymbol{x})$. Rule 2 then says that within $\boldsymbol{M}$, $\boldsymbol{N}$ may be exchanged for $\boldsymbol{x}$, yielding $(\lambda \boldsymbol{x}\boldsymbol{M}\boldsymbol{N})$.

Herewith a concrete example, one drawn from elementary algebra. (I am departing the formal system to offer the example.) Suppose that $M(x)$ has the form $2*x + 1$, and that $N(x) = 3$ for every value of x. The star * indicates multiplication, and I have included the symbol as a friendly gesture toward computer scientists. Application and abstraction thus yield

$$(\lambda xMx)Nx = (\lambda x.\ 2*x + 1)3 = 7,$$

the verbal music saying in this case that the specific function in which some number is first multiplied by 2, and then given an additional 1, when applied to 3, just happens to yield 7.

Rule 2 permits a further convenient compression of symbolism:

$$((\lambda x Mx)Nx) = (\lambda x.\ 2*N(x) + 1) = (\lambda x.\ 2*3 + 1),$$

or, what comes to the same thing,

$$(\lambda x Mx)Nx = M[N].$$

The third rule in effect reverses the second. If **N** has been substituted for ***x*** in **M** then the logician is allowed

3. To replace any part of **M** by ((λ***x*M**)**N**)—provided that ((λ***x*M**)**N**) is well-formed and the bound variables of **M** are distinct both from ***x*** and the free variables of **N**.

Changing a formula ***A*** to a formula ***B*** by means of these rules constitutes a *conversion* and as gravity is a basic force between material objects, conversion is a basic force in Church's calculus. In this respect, the calculus of lambda conversion is not a deductive system at all. There are no schemes of inference in the calculus, no axioms, and therefore no theorems. The calculus of lambda conversion plays over formulas and it contains only one modality of action and that is conversion itself, one formula giving way to another in a gelid but fluid exchange.

These rules are quite detailed; they are difficult to understand and wearisome to master. I would not suggest otherwise. But the rules have a deeper purpose and a more discriminating function. As Church himself remarks, "Rules [1–3] have the important property that they are effective or 'definite,' i.e., there is a means of always determining of any two formulas ***A*** and ***B*** whether ***A*** can be transformed into ***B*** by an application of one of the rules (and, if so, which one)."

An entirely abstract universe has been revealed to be under the control of a mechanical operation, and the juxtaposition of what is quintessentially human, and so accessible only to a human mind, and those mechanical rules by which energy is communicated from formula to formula is too odd and spans too great a defile to go unnoticed.

DOUBLE WORLDS

There is a duality at the heart of things that it is impossible to ignore and impossible to explain. Light is both a wave and a particle. Human beings are both spirit and matter. The shamanist's glyph of a double snake doubles again as a description of the molecular biologist's double-stranded DNA. Mathematical logic shares in the twofold mystery of things, the calculus of lambda-conversion now a set of symbols shuffled by the logician's stubby but unerring fingers, and now, as the light alters itself, an opening onto a world beyond the world of symbols. The formalism of the lambda calculus is intended to convey the motions palpable and permitted in a universe of functions, and so that sun is shedding its light onto a world in which there are infinitely many relationships, but hardly any things, just as women have long wished. It is a part of the magic of Church's apparatus that real and sparkling mathematical objects may be precipitated from those functions, another example of a mysterious doubling in the grand scheme of things. Functions retain their primacy in the system, but objects are defined *in terms of* functions, so that a single class of abstract entities performs virtually all the operations that would otherwise devolve upon still other mathematical objects, the operation rather resembling industrial production in the old Soviet Union, in which one standardized object would be forced to play innumerably many improbable roles: *iss tank, iss toothbrush, iss lawn mower, too.*

The natural numbers appear in Church's scheme by means of the following definitions. *Definitions,* note, in which particular formulas of the calculus have been picked out to play certain roles:

$$1 = \lambda f \lambda x(fx)$$
$$2 = \lambda f \lambda x(f(fx))$$
$$3 = \lambda f \lambda x(f(f(fx)))$$

and so on, farther up.*

These definitions lie beyond the margins of common sense. The number 1 has been identified as a function. But wherein lies the *number*? The identification of the number 1 with the set containing just the empty set (1 = {{∅}} has at least a certain visual appeal. Look inside. There is *one* thing there. Church's definition seems far less intuitive; at first glance—let us be candid—it seems incomprehensible.

And yet there is a core of common sense wriggling in the container of Church's definition. The natural numbers have a double hold on our imagination. They are in the first place things or things *like* things. To speak of the number five is to pick out *that* number from among the world's other numbers. It is this numerical sense that set theory captures efficiently.

But the natural numbers are also the overflow in mathematics of various human *acts*. A man lifts a hammer to strike a nail and thereafter the hammer bangs. An action has been undertaken. Prompted by the urge to do it again, he does it again. Having banged that nail five times, and after having also banged his thumb five times, he gives up. And this is one system of description, one way of describing what he did.

This system suggests a system in turn. Instead of saying that this lummox banged the nail five times, we might say that he banged the nail once and then repeated what he did. With the logician in command, repetition gives way to *iteration,* the

*I am departing somewhat from Church's own treatment; it doesn't matter.

repetitive act mirrored in iterative language: **Bang! Bang Bang; Bang, Bang, Bang; Bang, Bang, Bang, Bang; Bang, Bang, Bang, Bang, Bang.**

These terse bangs get the job done. They describe what took place. Parentheses compress the notation while lending it clarity:

Bang(Bang(Bang(Bang(Bang)))).

One thing has been done and then done again five times. The essentially iterative nature of the operation stands revealed—*start from the inside and just keep banging.*

Banging is, of course, beside the point. Iteration can easily play over arithmetic functions, as when $f(x) = x + 1$ is iterated five times:

$$f(f(f(f(f(x)))))),$$

yielding, when $x = 1$, the number 6.

Yet if banging is beside the point, so too $f(x) = x + 1$, which ties iteration to the very particular circumstances that the function $f(x) = x + 1$ has been iterated five times. It is iteration itself that counts. *Something* is being done to *something*. It hardly matters what is being done or to whom.

Alonzo Church may now be seen maneuvering himself upward into abstraction. Say that f is *any* function whatsoever and so any relationship. The five fold iteration of f

$$f(f(f(ff(x)))))$$

is an expression that has lost every tether to the ground of numbers and so stands naked in its essence, as Platonist might say. *And its essence is simply to iterate itself five times.*

An unexpected possibility now stands revealed. A number might be associated with the iteration of a function and such iterations might be designated from within the calculus of lambda conversion by certain formulas.

And this is precisely what Church's definition of the natural numbers achieves. An arbitrary function is selected, its inner arithmetical identity entirely a matter of indifference. The expression $\lambda x(fx)$ serves to designate that function and beyond this it serves no other purpose. Whatever it is, $\lambda x(fx)$ has done what it has done, operating on some variable x. The definition of the number 1 now follows: $\mathbf{1} = \lambda f \lambda x(fx)$. The expression $\lambda f \lambda x(fx)$ designates that function f operating on itself just once. But since $\lambda x(fx)$ serves simply to designate the function f, the number 1 is also equal to λff, or what comes to the same thing, $\mathbf{1} = \lambda ff$, using language almost identical to the language Church employed.

This definition represents a double triumph. Something has been made to answer to a natural number—a relationship of some sort acting on itself just once. The identity of the function hardly matters, questions of what it is displaced by an assertion of what it does—and what it does is to act on itself just once. At the same time, some formula has been made to designate that relationship—the formula λff.

Precisely the same pattern of justification holds for the other numbers, the numbers in general identified not in terms of any familiar objects, such as sets, but in terms of the concatenation of relationships. These definitions, although counter-intuitive, have the liberating effect of peeling from the very concept of a number everything save its essence.

Addition, division, multiplication, and subtraction enter into this world entirely as operations of functions performed on functions, but with the effect, which now begins to allow colors to seep through symbols, of re-creating (or revealing) the ordinary objects of computation—the natural numbers, incarnated in the Peano axioms and elsewhere in one guise, and incarnated again in Church's system in another guise. Given this double incarnation, it is easy, in fact, to peer down from Church's now soaring blimp and spot ordinary arithmetical relationships tootling on the ground.

The numbers 1 and 2 have already been defined in terms of

the calculus of conversion. If the definitions are to make any sense, it must surely follow that $1 + 1 = 2$, not only in the real world, where it does follow, but in the world where $\lambda f \lambda x(fx)$ and $\lambda f \lambda x(fx)$ must somehow fuse themselves to form $\lambda f \lambda x(f(fx))$.

Church's definition of addition trades on iteration, if only because there is absolutely nothing else upon which it could trade, and if the definition is astonishingly spare it is also astonishingly elegant. The sum of two numbers **M** and **N** is simply that function that has itself been iterated **M** + **N** times. Two men are now banging nails. The first bangs his nail five times, the second bangs his nail three times, the interminable racket of their banging proceeding in succession. They have, these two men, together banged two nails eight times, that sum easily imagined as an eight-fold repetition of some single primordial big bang.

The idea that captures addition amidst hammer and nails captures it as well in the calculus of lambda conversion. The metalanguage is now in force. I am talking about formulas down below and not wandering among them. If **M** and **N** happen to be the same number, their sum is a formula

$$\lambda f(\lambda f((\mathbf{M}f)((\mathbf{N}f)f))),$$

the symbols **M** and **N** serving to collect iterations on the function f.

With a descent downward from the metalanguage to the calculus itself, this general rule enforces the decision that common sense demands. It is **1** that is being added to itself. Very well. **M** and **N** are **1**.

Whence

$$\lambda f(\lambda f((\mathbf{1}f)((\mathbf{1}f)f))).$$

When the defined symbol **1** is replaced with the formula doing the defining, the displayed formula becomes an item within the lambda calculus itself. And plainly this formula is simply 2 = $\lambda f \lambda x(f(fx))$, or, what comes to the same thing, $\lambda fx(f(fx))$ or $\lambda ff.\ \lambda ff.$

Church invented the calculus of lambda conversion in order to articulate his own vision of a system whose operations proceeded by mechanical means. The most primitive of digital computers lay slumbering in the trap of time; programming languages were unknown. And yet, like so many of the great logicians, Church seemed to have had an uncanny feeling for the future, a sure sense that his elegant, difficult calculus might prove an instrument for machines that he could not imagine and an inspiration for men whose pursuits he could not have fathomed. In this, he was correct. The calculus of lambda conversion has been incarnated in any number of functional programming languages, its reticulated network of iterations no more difficult for a digital computer to handle than any other mechanical operation.

BINARY STARS

Certain stars in the cold depths of space appear to earth-bound astronomers as if they were linked, any sense of their real separation annihilated by their great distance from the earth. Very often they rotate about one another, turbulent plasma flashing between them, their separate identities irresolute so that at times they seem two stars and at other times one massively pulsing object.

Binary stars may be seen within mathematical logic as well as astrophysics, where they afford the logician the same quivering excitement as the astrophysicist. The lambda convertible functions comprise the class of functions that may be derived by means of lambda conversion from the well-formed formula of the calculus of lambda conversion. One class of functions and thus one star. The recursive functions are mathematical functions that may be derived from recursion's core by mechanical means. Another class of functions and thus another star.

Or so it would seem. Lambda convertible functions are abstract; recursive functions concrete. One can hardly imagine a connection between them. And yet the lambda convertible and the recursive functions are binary stars, circling around one another, fusing their identities, suggesting what no form of common sense could reveal, and that is a profound connection between two utterly distinct aspects of experience.

The recursive function $g(x) = x + 1$ takes a number and sends it to another number. If $x = 1$, then $g(1) = 2$. There is no simple way in which to translate this trifling affirmation into the lambda calculus; the calculus contains nothing to suggest that two items are equal.

If direct translation from one star to another fails, there is nonetheless a way in which $g(1) = 2$ may be *represented* within the calculus of lambda conversion. A part of the work has been done. The numbers 1 and 2 have already been given an interpretation in Church's system as the formulas λff and $\lambda ff.\ \lambda ff$.

Now suppose that some formula b is adjoined to λff. I am *within* the lambda calculus now and $b.\lambda ff$ is a well-formed formula *of* the lambda calculus.

An exchange of stellar material commences with a definition. The statement that $g(1) = 2$ admits of representation within Church's calculus if $\lambda ff.\ \lambda ff$ is *derivable* from $b.\lambda ff$ by lambda conversion. Conversion is an activity taking formulas to formulas; an ordinary function takes numbers to numbers. The association just forged links two activities by means of the logician's imperial command.

The definition I have just given covers only the case at hand. For a more general statement, ascent to the metalanguage is again required. Herewith a completely general statement. The function $f(m) = r$ is an ordinary arithmetic function, much like $g(x) = x + 1$, and m and r are the ordinary pussycats of algebra, standing for various numbers. Bold-faced symbols designate formulas within the lambda calculus. The function f is λ-definable just in case there is a formula **F** in the lambda calculus such that if $f(x) = y$ then **r** may be obtained from **F.m** by means of lambda conversion.

As in the case of $g(x) = x + 1$, conversion has been made to function as a surrogate in the calculus for the artifice of equality in ordinary algebra. The difference in definitions is a matter merely of generality.

Fusion is now about to take place. What Church (and the American logician Stephen Kleene) succeeded in demonstrating in 1936 was just that *every* recursive function is λ-definable and that moreover every λ-definable functions of the positive integers is recursive.

This is utterly astonishing. No other word will do. Mathematical logic has revealed that those far away and separate stars are revolving about one another and that contrary to every expectation, they are so inextricably fused that, shaking his head, the logician can only say that when all is said and done they are really revolving around a molten common core.

THE MONUMENT

After reaching retirement age at Princeton in 1967, Church decamped for the University of California at Los Angeles, the passage from one coast to another barely affecting his habits of work or the massive stability of his personal routine. From time to time, he would gaze over the contemporary philosophical scene, occasionally discharging his opinions in short, commanding, and surprisingly elegant little arguments. He continued to edit the *Journal of Symbolic Logic*; he continued his own research, his intelligence and grasp of logic so massive that it seemed to operate entirely without a cutting edge, the thing forging ahead like a sea-going oil tanker with a blunt square bow. I can only imagine him trudging through the warm sunshine of Los Angeles, his only concession to geographical dislocation involving nothing more radical than a reluctant shedding of his habitual Princeton overcoat.

He passed placidly through the rest of the 1960s, and then placidly through the 1970s and 1980s; one hardly thinks of the man as aging, but I am sure that he did. He worked uninterruptedly.

And then the curious obituary notice, in which time and the circumstances of his life placed him in Hudson, Ohio. A long illness, his son in attendance. The white clapboard midwestern house, with its broad front opening out to a lawn, baking in the summer heat. Fireflies in the evening and the smell of honeysuckle in the humid air, the large dense frame shrinking day by day, time having at last reached for the logician, just as I am sure he knew it would, the massive, orderly intelligence drifting upward like smoke disappearing in the summer sky.

CHAPTER 9

The Imaginary Machine

It happens and no one knows why. The intimation of an idea floats in the atmosphere for days and years and hours, and then at one time and at one place, the scattered parts of that intimation draw together into a thunderhead and drop their dense load of moisture onto the waiting earth.

The scattered parts of the thunderhead that broke with a clap in the 1930s can in retrospect seem destined to cohere—a concern for symbols and symbolism, the inferential rules of logic, the axioms for arithmetic, the idea of a universal language and so a universal calculating machine, the intuitive concept of effective calculability; these were things in the ambient air, but it was never quite certain whether they would form a coherent whole, or whether they would remain obstinately unproductive, like clouds that promise rain but then dissolve into nacreous wisps.

Kurt Gödel introduced the recursive functions into logical discourse; and Church, the machinery of the lambda calculus.

These were mathematical abstractions, their connection to the concept of an algorithm marked by a fairly long and fairly complex chain of definitions.

It was into the community of these abstractions that Alan Turing introduced an idea of an imaginary machine, a device which collapsed various abstractions into a single brilliantly simple construction; by attempting to catch himself in the act of thinking, he did what only genius allows anyone at all to do: he made something out of nothing.

SAD YOUNG MAN

The lives of the mathematicians tend to be imagined first by novelists and then lived by the mathematicians themselves. It would not surprise me to learn that Kurt Gödel was, in fact, a character first contemplated by Franz Kafka; as an American Gothic and so an American original, Alonzo Church was clearly the creation of Ayn Rand or John Dos Passos. Turing was English, and not American, but there is something in the peculiar nature of his personality, a form of longing, perhaps, that calls F. Scott Fitzgerald to mind, almost as if he were one of Fitzgerald's sad young men.

He had come to Princeton from Cambridge in 1936 in order to study mathematical logic with Alonzo Church, if only because Church at Princeton had already come to represent the subject's future mass and so like a magnet exerted an attraction as far afield as England, but however much rural Princeton afforded Turing the exciting opportunity of watching a subject in the process of creation, the association between Church and Turing must have been one of those curious affairs involving the misappropriation of two distinct styles—Turing, distant, high-strung, plainspoken, diffident, and thin; Church, remote, Gothic, imperious, and fat.

Turing had sandy hair dropping to a bland face in which only his lost lonely eyes caught and then reflected a somber reddish light, that face tapering unobtrusively into an ectomorph's body. Like so many unhappy misfits, he had fallen in love with long distances, loping for miles along the New Jersey back roads, smacking his forehead from time to time to strike an errant mosquito, the ground eaten up by his graceless but efficient stride. He was lonely, of course, separated from other men by the circumstances of his birth, accent, education, taste, and homosexuality.

There was no question, even from the first, that his was an unusual, even a singular, talent, in part mathematical, in part logical, and in part something distinctly other—that something other the reflection of an uncanny and almost unfailing ability to sift through the work of his time and in the sifting discern the outlines of something far simpler than the things that other men saw.

This was his gift, the one sure place that he could turn without effort and so without doubt.

THOUGHT REGARDING

A Turing machine is an imaginary object with real powers and properties, and if this is to suggest a paradox left loitering in the garden of mathematical logic, this is only because, like real toads in an imaginary garden, the paradox is a part of the picture and so a part of the pattern. It is, of course, precisely Turing's imaginary object that depicted in elegant outline the blueprint for a physical computer. Turing did witness the instantiation of his great idea in matter. A brilliant programmer, he then raised, as myth recounts, the question whether physical computers could think and so capture some elusive power of the human mind in matter. And if I say that this is myth, it is only to observe that a second blueprint lies beneath the first, like a rare oil discovered painted underneath a prim academic scene.

In almost all of his papers, Turing appealed primarily to the actions of what he called a "computer," meaning by the description, the actions undertaken by a *human* agent, one prepared (or forced) to manipulate a finite set of symbols according to fixed rules. Many men and women did, in fact, make a living as computers in the world that Turing knew: clerks, like Melville's scrivener, scribes of various sorts, accountants, tax collectors, notaries, recording secretaries, inventory managers, bankers, tellers, cashiers, draftsmen, an army of men and women rising early and then spending the hours that followed at wooden desks, where they would add columns of figures or copy briefs or annotate blueprints or otherwise carry on with an activity that was as disciplined as it was dispiriting.

What Turing imagined, as he tried to catch himself thinking, was an impalpable machine that could do what human computers did, the act of imagination striking and utterly original if only because Turing was not interested in any particular mental act, such as might be involved in executing the arithmetical functions, but in the essentials of human computation itself. It is this relentless drive for the fundamentals that led Turing to the deep and very disturbing conjecture that insofar as human beings are engaged in thought *of any form,* they are acting as human computers, the difference between a woman settling her weary haunches on a cashier's stool in order to punch numbers into a primitive calculating machine and a woman imagining and then recounting the infinitely amusing activities taking place at Mansfield Park, a matter in the end of detail, self-consciousness, and the small embroideries of inessential differences.

The blueprint for a Turing machine is both architectural and procedural. The architecture describes the machine's four parts, and these are common to all Turing machines; the procedures comprise its instructions, and although they are all written in the same code and obey the same format, they vary from machine to machine.

Architecture

An infinitely long tape: The tape is divided into squares and stretches in two directions.

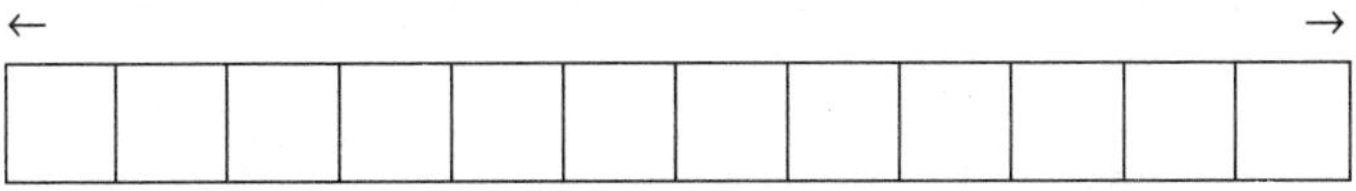

← →

The fact that the tape is infinitely long means that a Turing machine has an infinite amount of memory; whatever the nature of its calculations, it is not apt to be embarrassed by a shortage of space.

A finite set of symbols: A Turing machine is intended to execute absolutely simple, primitive operations. Its symbolic requirements are very modest. Only two symbols are needed: 0 and 1. This is something that Leibniz would have understood; it is something he did understand.

A reading head: A reading head has the triple power to (1) scan squares one at a time, (2) move to the left or to the right one square at a time or not move at all, and (3) inscribe or erase symbols on squares. The reading head is thus endowed with movement, a steady gaze, and the editor's enduring itch to make everything so much better by rewriting what it spots or sees.

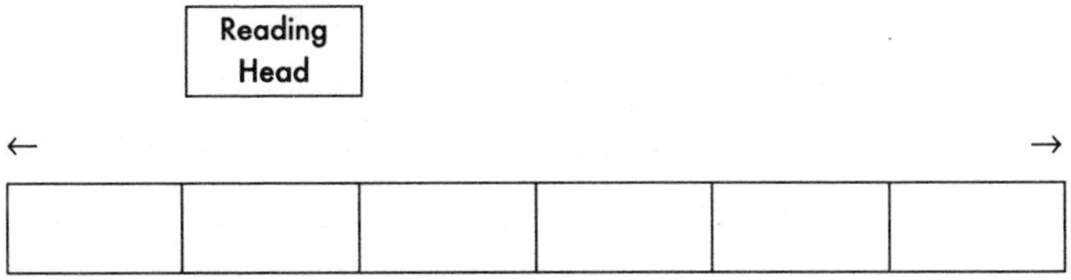

Reading Head

← →

A finite set of states: The states correspond to various finite internal configurations of a Turing machine's reading head. Beyond saying that the reading head *has* a finite number of parts and so can be configured and reconfigured into various finite arrangements or states, the logician has already coolly declined the reader's invitation to specify those parts in any way.

What has been given, then, is an infinite tape, a set of symbols, a triple-powered reading head, and a set of states. A Turing machine involves a very daring act of simplification.

Procedures

The behavior of a Turing machine is controlled by a finite series of instructions.

In the nature of things, instructions are always governed by their content and the circumstances under which they are made, the command *Jump*! making sense if addressed to a man or a toad, but not making sense at all if addressed to a church steeple or a paramecium.

In the case of a Turing machine, there are only two relevant circumstances. One is the state of the reading head; the other, the symbol that it is scanning.

By the same token, there are only two commands intelligible to a Turing machine. The first instructs the machine what to write or erase; the second, whether it is to move to the left or to the right by one square.

This means that each command given to a Turing machine is in four parts, the parts mediated by a hypothetical:

> *If the machine is in this state, and is scanning that symbol, then it must write this symbol or that symbol, and move one square to the left or to the right or not move at all.*

State and symbol comprise the circumstances under which the machine acts; writing and movement, the content of its action.

It follows that the entire behavior of a Turing machine can be controlled and so governed by a set of instructions in *if-then* form, one whose antecedent specifies the circumstances of action and whose consequent specifies its content.

Having been devised by Turing in the mid-1930s, the Turing machine made possible because it made palpable the design and then the construction of the first all purpose digital computers, Turing's imaginary machine containing within itself all of the clues—input, output, states, fixed program—needed to fashion a concrete machine from its conceptual model. The advances in technology required to actually build a digital computer were altogether more prosaic, of course—a growing mastery of electronics, the development of transistors and integrated chips, a new-found familiarity with rare metals and their obscure properties, a willingness to tinker and to take chances, and luck. It is not entirely clear whether the Turing machine brought its ancillary technology into existence, or whether that technology would have muscled its way on the scene regardless, but whatever the ultimate metaphysics of the matter, Turing's idea played some role in the coordination of events and so represented once again the power of pure thought to shape matter to its ends.

ACTION

If a Turing machine is radically simple in its design, it is radically simple in its action as well; it is indeed so simple that it is hard to imagine the device capable of doing anything at all. It is remarkable, then, that in a certain sense a Turing machine is capable of executing virtually any tightly specified intellectual act, its alarming competence apparently at odds with its enviable simplicity.

This is a large and an astonishing claim, perhaps the largest and most astonishing claim of the century; but before the claim can quite be countenanced, the reader with his or her tense thumb on the margins of this page may wish for a demonstration of the more modest claim that there is *something* that a Turing machine can do.

What follows is the description of a Turing machine that can add any two natural numbers. *Any* two, note. The capacity of the machine is unbounded and in this it resembles the human mind more than it resembles the physical computers to which it gave rise.

This machine is entirely standard—straight off the production line, in fact. It comes with a glistening infinite tape, a reading head divided into states, a single symbol, 1, and a ten-part program. The machine represents a natural number n to itself as a sequence of $n + 1$ consecutive 1s, so that 0 is 1, and 1, 11, and n, $n + 1$ occurrences of the numeral 1. The sum of two natural numbers $m + 1$ and $n + 1$ is thus $(m + n) + 1$.

A Turing machine for addition comes to its labors with symbols *already* inscribed on its tape; its business, after all, is solving problems and not setting them. *I* have asked a Turing machine to add one to itself, offering it a tape whose otherwise unbroken expanse is punctuated by four symbols:

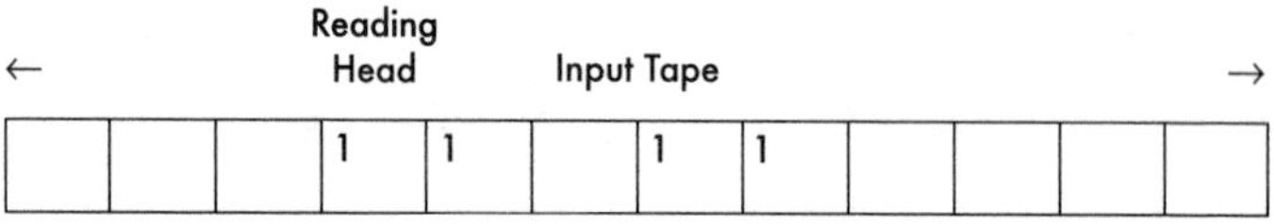

← Reading Head Input Tape →

			1	1		1	1				

The blank between symbols indicates that the numbers on eithe side are to be added and this, too, is a part of our understanding (I am involving the reader in a collusive intellectual act that is not in any way reflected in the machine's architecture.) The machine's reading head is positioned directly over the left-most symbol. It fiddles with the symbols somehow and when it is through fiddling, what it has been given emerges as what it has gotten. Where before there were two occurrences of the symbol designating one, afterwards there is one occurrence of the symbol designating two.

The program is simple. Save for the fact that an imaginary machine is now doing what for time immemorial only human beings could have done, there are no surprises. The human observer now looking over my shoulder has no doubt already spotted the pertinent algebraic fact that the numbers $m + 1$ and $n + 1$ transform themselves into $(m + n) + 1$ just in case 1 is added to the first number and 2 subtracted from the second—$(m+1) + 1 + (n+1) - 2 = (m + n) + 1$. The machine's code is thus devoted to getting the machine to add one symbol to the tape and to delete two others.

Each line of the program is divided into two parts. The first part describes the symbol that it is scanning and the state in which it finds itself; the second part describes what the machine is to do and the state into which it is to move. The symbol B indicates that a square is blank. The action of this machine quite nicely subordinates itself to a familiar checklist:

State and Symbol	What it does and where it goes	
1) State 1, 1	Moves right, stays in State 1	Check
2) State 1, B	Prints 1, goes to State 2	Check
3) State 2, 1	Moves right, stays in State 2	Check
4) State 2, B	Moves left, goes to State 3	Check
5) State 3, 1	Erases 1, stays in State 3	Check
6) State 3, B	Moves left, goes to State 4	Check
7) State 4, 1	Erases 1, stays in State 4	Check
8) State 4, B	Moves left, goes to State 5	Check
9) State 5, 1	Moves left, stays in State 5	Check
10) State 5, B	HALT	Check

The checklist and the instructions that it reflects have an obvious internal coherence. The machine moves rightward until it discovers a blank space, whereupon it prints 1, thus in effect adding 1 to *m*. Promoted to State 2, it moves rightward again until it again discovers a blank. The rightward blank is its signal to change states and turn around. Thereafter it erases the next two 1s and leaves them blank, thus subtracting 2 from *n*. It then marches leftward until it uncovers the first left blank and then draws its meditations to a HALT.

When the machine has finished up its work, there are three contiguous inscriptions of the numeral 1 where before there were four separated inscriptions of the same numeral. As expected the machine has determined that 1 + 1 = 2, an achievement that it represents to itself as (11) + (11) = (111).

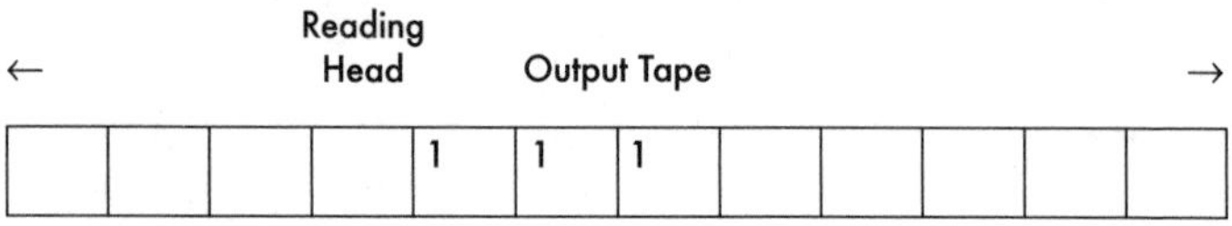

← Reading Head Output Tape →

				1	1	1					

The reader inclined to object that with respect to 1 + 1 = 2, she knew it all along has, of course, missed the subtle signs of the miraculous in this demonstration, for the very same set of instructions establishing the sum of one and one are capable of establishing the sum of *any* two numbers of whatever size, ten lines of simple code sufficient to bring the infinite under finite control.

A LIGHTHOUSE OF SORTS

A Turing machine divides the present from the past in the decisive and unalterable way in which all great inventions or ideas split the continuum of time—it is now impossible to imagine the world without it, just as it is impossible to imagine the world without Hamlet or Newtonian mechanics. The causal nexus that has made the modern world extends in a simple line from Turing's ideas directly to the ever-present, always-moving now. No less than the genetic instructions that are said to pass through the generations, certain ideas are capable of penetrating the future.

If Turing's imaginary machine has played an important role in the development of technology, it has played a still more important role in the history of thought, providing a simple, vivid, and altogether compelling mathematical model of the ancient idea of an algorithm. Whatever it is that a Turing machine does, it does it effectively by getting the job done, if it can be done at all.

At first, this might seem far from the community of concerns embraced by Gödel and Church. The appearance of distance is misleading. Gödel and Church looked toward certain functions in order to specify the idea of an effective calculation and so to give content to the concept of an algorithm. A Turing machine takes symbolic inputs to symbolic outputs; the conversion of one set of symbols into another set of symbols comprising the formal record of a transformation that may perfectly well express an ordinary mathematical function.

The Turing machine whose program allows it to convert 11 to 1111 and 111 to 11111 and 1111 to 111111 has in its calculations expressed the function $f(x) = x + 2$. The Turing machine whose program I have already outlined has busied itself calculating the function $f(x, y) = x + y$. The connection between what a Turing machine does and the functions that it computes is clear enough to prompt a definition. An ordinary arithmetical function is Turing computable just in case there is a Turing machine that accepts its arguments as inputs and delivers its values as outputs.

With this definition given, a Turing machine now takes its place among the primitive recursive and the lambda-convertible functions, since it is intended to realize precisely the class of functions that submit themselves to the authority of an algorithm. A third definition of an essential concept has been introduced. The algorithm has made its advent in terms of the primitive recursive functions; it has made its advent again in terms of the lambda-convertible functions; and again in terms of the Turing computable functions.

Three definitions, three representations, but only one idea.

A Turing machine is an abstract mathematical object, something that belongs to the same conceptual category as sets or groups or rings or ideals in the severity of its logical profile; but a Turing machine is also a device that *does* something and so belongs to the instruments of action, the thing, against all expectations, coming to take its place in a world where the stream of time is diverted by means of agency, intention, and thought. There is in all this another curious reversal of pattern, as when velvet is brushed backward. The aim of Western mathematical science has always been to gain access to the divine view of creation, physicists arguing broadly that the universe can be grasped and so understood only in terms of its secret mathematical laws, the deity evidently having amused himself since before time began by laying out the architecture of things in his own irritating and inscrutable way. Science as a great quest will have reached its goal when physicists discern the single set of luminous laws by which everything can be explained. It is a chilly and a static vision, timeless in its austerity, the quest itself, as laymen have long sensed, involving a certain profound derangement of the sensibilities. A Turing machine is not a law of nature, but a thing: it may be used, but it does not—it cannot—explain; and in this sense, its creation marks the first step in a process of historical detachment.

ENIGMA

Let me tell you about the very talented. They are different from you and me. They possess and enjoy early, and it does something to them, makes them soft where we are hard, and cynical where we are trustful, in a way that, unless you were born talented, is very difficult to understand. They think, deep in their hearts, that they are better than we are because we had to discover the compensations and refuges of life for ourselves. Even when they enter deep into our world or sink below us, they still think they are better than we are. They are different.

Alan Turing was very, very talented, and so very, very different; the talent, and so the difference, is reflected in the radical way in which he changed the landscape of logical thought: where before his work there were only the looming cloud-shrouded, thoroughly inaccessible peaks created by Gödel and Church, after his work, there is an opening in the landscape, the cold clouds disappearing and the mountain peaks giving way into a fragrant valley of fruit trees in bloom. Gödel's incompleteness theorem is a very hard climb, and we who look at it with admiration also look upon it with dismay, but Turing's machine is easy to understand and it explains itself, the goal of explicating effective calculability not only realized in his machine but realized as well in our appreciation of his machine, the effect, akin to a hat trick, one in which simplicity of execution is twice on display, magical in its ease, and this talent of his, to reorganize what is difficult and to make it seem both comprehensible and inevitable, reflected the calm assurances of his deepest nature, which was to take what was complex and see in it what was simple and so what was important.

Turing completed his great work in logic in the late 1930s. Thereafter, he returned to England. There follows a story at once curious and strange, one that again reveals the oddity of the man, his capacity to refine the world to its essentials.

The world is now at war, booming armies cutting their swath through Europe and Asia, all the sad young men who had collected themselves at Princeton or at Cambridge in order to think all their sad young thoughts, now dressed in khaki, tramping along dusty parade grounds and taking orders from tough sergeants who knew absolutely nothing of mathematical logic or English literature or Renaissance poetry.

You and I who trade in ideas carry on in a world in which everything is valued more than the commodities that we exchange; but a world at war is a world in which quite suddenly intelligence is valued, not by tough sergeants but by the politicians who give them orders, men who in peacetime would never think to stand us to a drink quite suddenly reaching over to palpate our shoulders while signaling to the bartender in their easy and familiar way.

And so, Alan Turing, who had spent his time at Princeton allowing his thoughts to grow up beautifully, found himself in a world where armies clash, his new masters indifferent to his dreams but eager to use and so exploit the machinery of his dreamworks. He was no longer a very young man, but he was still young—twenty-eight in 1940—still half-formed in the way in which talented men never quite finish filling themselves up, and this radical change in circumstance in which for the first time powerful and determined men were willing to overlook his oddity must have seemed to him further evidence that his talent was a form of protection, a kind of insulation that would work in any weather, hot or cold.

He was recruited to work in military intelligence. Now the curious thing about military intelligence is that it must be both transparent and opaque, for if the intelligence is transparent it is no longer intelligence; and if opaque, it is no longer useful. The men who gather themselves into cramped and airless rooms and who send memoranda to one another marked Top Secret go to immense trouble to devise methods of encryption that lie on the very narrow border marking what is useful to one's own side and useless to the enemy, and it is one of the ironies of the story I am telling, that during the time that Alan Turing was loping along the alien roads of rural New Jersey, he never imagined that with a turn of the wheel of fortune, he might place his vivid dream of embodying human calculations in the confines of a machine in the service of some immense practicality, one in which ships might be saved at sea or cities saved from aerial bombardment.

By 1941, the Germans were the masters of Europe; directing their armies over a far-flung empire, the German High Command reposed much of their confidence in a cipher machine dubbed Enigma. Looking much like an elaborate typewriter, the machine was composed of a series of ratcheted cylinders, so that each letter of the keyboard, when depressed, could be encoded by hundreds of possible ciphers. In 1928, the Polish cypher bureau, BS4, had an opportunity to study an original Enigma machine and were able to build a model which they provided to British intelligence. And therein lay a tantalizing problem. British intelligence now had the ability to detect coded German messages by intercept. (So could everyone.) What they lacked was the encryption key between the machine and those messages, the one that would reverse the effect of coding and render the message transparent.

It is a combinatorial problem of enormous complexity, requiring not so much insight as ingenuity for its solution. The Germans changed their encryption key daily and so the solution to the problem, if it was to do any good at all, needed to be available in real time. You and I can feel the full force of the problem by means of a simple question: given a string of gibberish known to make sense and a cipher machine, read the gibberish quickly enough to act upon the information that it contains; but *we* must depend on others for its solution, the outer circle, beyond the very talented, consisting of men who must know something and the inner circle consisting of men who can tell them what they need to now.

The British tackled this task with an enterprise born of desperation. A team of code breakers and mathematicians was assembled at Bletchley Park. Many of the details of their work are still secret; but within months, Alan Turing, working largely in isolation, had solved the essential cryptographic problem. Thereafter, coded communications sent by the German High Command to their armies were accessible to British intelligence. When Winston Churchill learned by this means that Coventry was to be bombed, he was forced to allow the bombing to proceed, lest the Germans divine that their code had been cracked, an interesting example of intelligence defeating itself, but in other theaters of war, especially in the North Atlantic, the information that Turing was able to decipher played an all-important role, time and again, British attack planes appearing in the most mysterious way directly above surfacing Nazi U-boats.

Our story is concerned not with a war but with a man and the way in which his talent made him different from other men; but after a point, the story trails off, in the way in which all stories about the very rich or the very talented trail off.

Turing retired from his wartime duties; he joined the staff of the National Physical Laboratory in London, leading a team in the design of an Automatic Computing Engine. Thereafter, he became deputy director of the Computing Laboratory at the University of Manchester. The great and warming fire of his talent remained undimmed. Like von Neumann in the United States, he moved easily between abstract thought and concrete details, developing new programming techniques and feeling his way forward from what a computer might be to what a computer would be. He wrote a number of provocative papers on morphogenesis; and he wrote an historically important paper on artificial intelligence in which he argued slyly that a computer capable of fooling a human interlocutor must be considered intelligent just because it *could* fool a human interlocutor.

He disdained to conceal his homosexuality, imagining, perhaps, that his own sexuality as an adult would fall behind the sturdy defenses of his talent and so come to be treated by other men as an indiscretion or an idiosyncrasy, something personal and so something harmless. In this he was mistaken. He never quite understood that like all men marked by a high and unusual talent, he lived in a world of enemies, and by proudly failing to conceal his homosexuality, he opened the fatal chink in the armor that his talent provided and he learned too late that without his armor he was destined to live like other men and so was doomed by his destiny.

The story grows murky. Officials of the British government, concerned perhaps over lapses in security, forced Turing to accept a course of hormone therapy designed to suppress his libido. His skin grew smooth; his voice changed its register. The preposterous course of treatment served only to depress and demoralize the man; he became despondent, and with his own great talent shown at last to be a gift and not a way of life, sometime on June 7, 1954, he ate an apple laced with cyanide and died alone, the apple half finished lying by his bed.

There is one last ray of light that shines out from our story. Whatever practical work he may have done, Alan Turing was first and he was foremost a great logician, his deepest talents understood only by the men who had shared that talent and knew in the end what it meant. The world knew that he died, the logicians, *why* he had died, Gödel remarking that perhaps Turing had been unhappy because he wanted to get married but could not; and while this remark indicates either an astonishing degree of innocence, or the reverse, it does contain a portion of the truth inasmuch as Turing placed his confidence in being different from other men and discovered too late that except for his talent, he was no different at all.

And that's enough, Scott. That is quite enough.

CHAPTER 10

Postscript

To the uninitiated, logic often seems the driest of all disciplines, its formulas and the strange scrupulousness that it demands a rebuke to spontaneity and so to freedom. I stress that this is how the subject *seems*; in fact, as in all of mathematics, a river of real life rushes just below its surface, and the passions that it engages are human passions, the more so than in other branches of mathematics inasmuch as logic is itself inextricably tied to the loom of language and what language can or cannot do. No one studying the lives of the great logicians can miss the aching connection between what they were and what they did. Gödel consumed some dense spiritual substance in order to obtain the nervous energy necessary to achieve his results, and throughout the 1930s, he was forced repeatedly to repair to various sanatoriums in order to regain his sanity. Church lived his life as a part of his own calculus, subordinating himself to his symbols. Turing lived in the loneliness of his radical sense of simplicity, accepting in the end that it was simpler to die than to live. And when one ticks through the subject during the days in which mathematical logic was coming into creation as an esoteric, powerful, and disturbing discipline, the men who made the subject, almost without exception, reflected in their own personalities the pain of their passion, some collapsing in nervous exhaustion, others going mad entirely, still, others such as Alfred Tarski armoring themselves in personalities shielded in steel, and others still, finding a refuge of sorts in drink or in drugs.

The articulation of the algorithm took place within a span of a few short years, logicians in different parts of the world rooting through the same conceptual landscape; and if Gödel, Church, and Turing reflect nothing so much as the compressive ability of genius, the story is large enough to encompass other talents and so other men—Emil Post, for example, who took his place in a competitive arena of insanely high standards and persevered against all obstacles, his mind vigorous, alert, disciplined, and inventive, but his fundamental talent circumscribed by the fact that he lacked Gödel's supreme combination of elegance and power, or Church's monumentality, or Turing's sense of radical simplicity. He was gifted but not great, but he managed his gift with such care that in retrospect, it hardly seems to matter at all.

Emil Post was born in Poland in 1897 and came to New York as a part of the great wave of eastern European Jewry that moved sluggishly across the European continent and then crossed the Atlantic Ocean. He had lost his arm as a very young man and so faced the world both as a cripple and a foreigner. With his family settled in New York, Post participated in that curious but characteristic generational exchange in which his talent and hard work served to redeem the sacrifices of his parents. He was educated at Townsend Harris High School and at Columbia University, and there is some dense crabbed energy to everything that he did in mathematics that reflects the peculiarly intense rhythms of the city itself; he suffered throughout his life from manic-depressive disorders, the mania brought about cruelly by intellectual excitement, and the depression brought about by the mania, but somehow, setting Post the man against the New York of the first half of this century—*my* New York, as it happens—it is plain that he in some way absorbed the flickering and flashing of the city as it rumbled about him, his own thoughts now like the IND express that rattles relentlessly down Broadway, passing from 125th Street to 59th Street in a great indignant roar, the local stations whooshing past, and then afterward like Sixth Avenue on a rainy Sunday evening, gray and gloomy and unmoving, the smell of corned beef and beer issuing from the few taverns still open.

He came to mathematical maturity under the influence of Russell and Whitehead's *Principia Mathematica,* and his early work in logic, undertaken while at Princeton University, has an unmistakably prophetic air, for in essence, Post had seen the facts of incompleteness in 1920, a full ten years before Gödel published his epochal paper; but what Post saw, he saw in a jumble, the lines of inference clouded and often confused, and the blazing brilliant proof that forced these various lines to converge was simply beyond his reach and so beyond his grasp.

In 1941, when the signal achievements of the great men had been noted and recorded, Post wrote an inexpressibly poignant letter to Hermann Weyl, then editor of the *American Journal of Mathematics:* "It is with some trepidation," he began, "that I submit the accompanying paper, 'Absolutely Unsolvable Problems and Relatively Undecidable Propositions, Account of an Anticipation,' for publication in the AJM." In the paragraphs that followed, he asked only that his anticipation be noted for what it was.

Weyl's response was a model of brutal efficiency: "I have little doubt," he said, "that twenty years ago your work, partly because of its then revolutionary character, did not find its due recognition. However, we cannot turn the clock back; in the meantime Gödel, Church and others have done what they have done, and the *American Journal* is no place for historical accounts. . . ."

But then Weyl, realizing what he had just said, softens: "Personally," he adds, allowing the logician's mask to slip just once, "you may be comforted by the certainty that most of the leading logicians, at least in this country, know in a general way of your anticipation."

The human drama that these words reveal, in which the shadow Post threw is obscured by the larger shadows thrown by other men, is not the whole of the story and so not the whole of Post's life; he did immensely important work in logic that is quite genuinely his own, the strong engine of his industry continuing to hum and chug throughout the 1940s and early 1950s. His work on what have come to be called Post production systems profoundly influenced the course of modern linguistics. He had queer anticipatory powers and while he was incapable of realizing some of his visions because other men realized them more completely, he was granted other visions at other times, and some of these he realized completely and made his own.

Living as every logician did in the imperious court created by David Hilbert, Post was concerned during the 1930s to give form and content to the idea of a mechanical procedure. His line of attack was very much like Turing's: he thought in terms of something occupying a place in the imagination, and in retrospect it is sometimes said that Turing and Post hit on the same idea at the same time, the fact that Turing continues to gather the credit for what is a double invention further evidence that Post was doomed to live his life in the shadows thrown by larger men.

But this is not quite accurate. There is a subtle difference between the two men, and the penetration of the future that they both devised proceeded along different routes and by different means. Like Turing, Post imagined an infinite tape divided into squares, and like Turing again, he thought in terms of a space of symbols; but Turing placed a reading head above the squares, endowing the head with a finite number of internal states. That reading head is gone from Post's machine and instead there is a human computer pressed into service, one entirely compelled to act in accordance with a certain finite list of directions.

That human computer, or "worker," as Post describes him, is capable of performing the following acts, which comprise the whole of his mental endowment:

(a) Marking the box he is in (assumed empty),
(b) Erasing the mark in the box he is in (assumed marked),
(c) Moving to the box on his right,
(d) Moving to the box on his left,
(e) Determining whether the box he is in is or is not marked.

In addition, that worker is equipped with an inner ear allowing him to follow certain highly formalized directions. These have the effect of prompting the worker. There is an overall directive:

Start at the starting point, and then follow directions;

those directions, in turn, compelling the worker either to

(A) Perform operations (a), (b), (c), or (d), or to
(B) Perform operation (e), whereupon he is once again to
(C) Perform operations (a), (b), (c), or (d) or to
(D) Stop.

Now this is much like a Turing machine, with the directions playing the role of states; and in fact, the difference between the two concepts is marginal.

But there is a curious point of emphasis in Post's machine that deserves comment. The Turing machine is an idealized computer, something *we* can see in retrospect; and there is a nice division in Turing's machine between hardware and software. The tape and the reading head represent hardware in prospect—this quite literally; while the instruction set represents software. Post's machine is entirely symbolic. Hardware has dwindled to a vanishing point. That poor drone, the human worker, is there only to follow instructions. In this respect, Post has fashioned an anticipation not so much of a digital computer as of its *program*.

And yet a Turing machine and a Post machine are alike in power. This may suggest that the distinction between hardware and software is to some extent an illusion. A machine as it is ordinarily understood belongs to the world of matter and follows the laws that regulate matter in its many modes. The machines devised by Post and Turing do have an instantiation in a world of plastic, silicon chips, copper wires, and electrical circuits; and within the ambit of the contemporary world, the computer seems as much a machine as an automobile or a lathe. And yet, I am tempted to say, this description represents a displacement of emphasis, a distortion in tone. Post and Turing have created machines responsive to a world of *thought*, and not matter at all, and the fact that when these machines are realized they must be realized as material objects represents nothing so much as the inevitable concession that mind makes to matter in a world in which it is only matter that has a kind of enduring stability. The essence of their machines is elsewhere, in a universe in which symbols are driven by symbols according to rules that are themselves expressed in symbols.

The place in which these machines reside is the human mind.

On the testimony of his students, Emil Post was a remarkable teacher, both demanding and exact. And he was a superbly disciplined scholar, working with determination under difficult conditions—no secretarial help, no real office, endless teaching responsibilities, and at home, little space and a young daughter underfoot, doing her best to heed the house rules that when her father worked in the living room, she herself must remain still. The manic-depressive disorder from which he suffered could not in the first half of this century be brought under the control of drugs and so Post was able to gain some measure of ascendancy over his affliction by means of sensible work habits, careful attention to diet and sleep, and the restorative of long walks. He measured his research activity in minutes, inscribing his thoughts meticulously in small leather-bound journals, and when the time allotted to research was up, he refused to press his ideas to their conclusions lest he perturb the delicate balance of his sensibilities and crash into frank mania.

On those occasions when his affliction overwhelmed him, electroshock therapy was the only treatment modality offering the promise of relief, a burst of electrical energy evidently serving to descramble his nervous agitation, much as a single loud shout can sometimes still a number of querulous voices. When in 1954, the mania grew out of control, Post repaired again to this hideous but strangely effective treatment; but this time the shock did more than quiet his agitation. He died of a massive heart attack directly after treatment, the electroshock stilling his life along with his mania.

A photograph of Post survives from the early 1950s. Post is sitting; he is somewhat slumped. His wife and daughter are on either side of him. They are both handsome alert women. And both women, although facing the camera, convey an extraordinary physical solicitude, a sense of tenderness. The photograph is a study in tension. It reveals, as so many of these documents do, something that goes beyond the tenderness. It reveals that this man is at the very center of both of their lives. It reveals that both women love him profoundly. It reveals that he is irreplaceable. And it reveals something more.

It reveals that they are terrified.

CHAPTER 11

The Peacock of Reason

The idea of an algorithm had been resident in the consciousness of the world's mathematicians at least since the seventeenth century; and now, in the third decade of the twentieth century, an idea lacking precise explication was endowed with four different definitions, rather as if an attractive but odd woman were to receive four different proposals of marriage where previously she had received none. The four quite different definitions, it is worthwhile to recall, were provided by Gödel, Church, Turing, and Post. Gödel had written of a certain class of functions; Church, of a calculus of conversion; and Turing and Post had both imagined machines capable of manipulating symbols drawn from a finite alphabet. What gives this story its dramatic unity is the fact that by the end of the decade it had become clear to the small coterie of competent logicians that the definitions were, in fact, equivalent in the sense that they defined one concept by means of four verbal flourishes. Gödel's recursive functions were precisely those functions that could be realized by lambda-conversion; and the operations performed by those functions were precisely those that could be executed by a Turing machine or a Post machine. These equivalencies, logicians were able first to imagine and then to demonstrate.

And by the time this was realized, the algorithm had at last made its advent.

CHURCH TENDERS A CONJECTURE

The fact that one concept had been defined by four definitions seemed to logicians of the 1930s an enormously suggestive circumstance. Many logical concepts are tied very specifically to certain systems of notations. What counts as a valid inference depends, for example, on the formal system in which the inference is embedded. *If Miss Blonde World is blonde, then someone is blonde.* No doubt. And yet if inference is restricted to the propositional calculus, which countenances only connections between completed sentences, *If Miss Blonde World is blonde, then someone is blonde,* is swallowed into a single sentence, the tantalizing connection between Miss Blonde World and the color of her hair disappearing altogether.

The concept of an algorithm seemed altogether otherwise: it remained the same regardless of definition, the equivalencies between various definitions all the more striking inasmuch as the definitions were so very different. A concept indifferent to the details of its formulation, Gödel asserted, is *absolute*. And in commenting on the concept to an audience of logicians, he remarked that the fact that only one concept had emerged from four definitions was something of an epistemological "miracle." It is an odd word for a logician to use, but the sense of the miraculous that Gödel invoked represents the odd place that the concept of an algorithm occupies in the catalog of mathematical concepts. An algorithm is a perfectly well defined mathematical object; but it is as well a human artifact, and so an expression of human needs in a way in which the derivative of a real-valued function or a Fuchsian group is not. Mathematics has always claimed for itself the utterly mysterious power to create concepts in alignment with the real world. In the case of the algorithm, that mysterious and controlling power has now been retracted from the real world to the world of the imagination, so that for the first time, the age-old human ability to count and to calculate and to reckon finds itself trapped within the silken confines of the thoughts that it itself brought into being.

Wherever human beings have sought to make for a connection between desire and action, they have turned to formal practices, legal codes, codicils, recipes, love letters, Books of Prayer, war manuals, tax tables—all the varied instruments which coordinate the flow of action from moment to moment, and so to spread the net of human consciousness beyond the episodic.

There is, nonetheless, a considerable difference between the purely human concept of effectively getting something done and the purely mathematical concept of computability, as it came to be defined by Gödel, Church, Turing, and Post. The human concept is larger, and it is imprecise.

In a paper of great historical importance, Alonzo Church attempted to pull the various threads of these observations into the knot of a single dramatic conjecture, one that forces a curious concordance between two sides of an idea. On the one side, there is the intuitive notion of effective calculability, the class of activities for which there is a corresponding but informal algorithm. On the other side, the recursive functions or the calculus of lambda-conversion or the machines devised by

Turing and Post. The concordance that Church proposed is simple. Insofar as mathematics can express an informal concept, it does so completely by means of any one of these four equivalent mathematical definitions. For reasons of clarity and intuitive appeal, the Turing machine has come to seem aesthetically best suited to represent the mathematician's particular vision; and so that concordance has an even simpler formulation: *Whatever can be done effectively can be done by a Turing machine.*

Such is Church's thesis. It is itself not demonstrable, of course, inasmuch as it associates in a single figure two quite different kinds of experience, the one human and informal, the other mathematical and precise. Post regarded it as a law of nature, no different in kind than Newton's law of universal gravitational attraction, but laws of nature are themselves expressed as relationships between mathematical quantities; Church's thesis draws a relationship between something that is not mathematical and something that is. To the extent that logicians have accepted Church's thesis, they have found evidence for it at every turn, the prophecy that it makes at least partially self-fulfilling.

Whatever its status, Church's thesis has had a powerful, if constraining, hold on the imagination of logicians, philosophers, and psychologists. And for obvious reasons. A great many human activities fall under the rubric of effective calculability: adding, dividing, subtracting, and multiplying the natural numbers are the obvious textbook examples, but what of deciding, planning, speaking a natural language, listening, understanding, plotting, devising, creating, entertaining, avoiding, seeing, spotting, remembering, advising, construing?

For that matter, what of falling in love, or figuring out taxes, or swimming the five-hundred-meter medley, or reading a textbook, or making a purchase, or devising legislation, or cooking a meal, or excising a mole, or conducting a political campaign and running for office, or washing the dead?

A great many things that we do *seem* to break the flow of time into discrete segments. And since these operations are human, they *seem* to fall under the control of a general symbolic apparatus, the algorithm appearing now as the quintessential device that human beings use for the mastery of time.

THE FUNCTION G— THE INSCRUTABLE

■ ■ ■ ■ Although a Turing machine was designed to *do* something and so shares the nature of all devices that perform a constructive task, the task it performed best is a task it could not perform at all; and if this is to suggest a place in which the ribbon of thought is pulling against itself, this is only because the history of logic is very much the creation of Juan Luis Borges, who was vouchsafed the singular ability to write stories that he never wrote.

In 1928, the German mathematician David Hilbert, conveying his ideas to other mathematicians at a Congress held in Bologna, expressed the hope that with the advent of the algorithm, mathematicians might discover a decision procedure for all of mathematics (and thus for all of life) so that coming upon a new conjecture, they could determine mechanically, and with complete certainty, whether it was true or not.

Hilbert's ambitions have remained unfulfilled, even though some of his followers continue to carry on his work, apparently unwilling entirely to accept the fact, long since demonstrated, that his ambitions have remained unfulfilled because they are *unfulfillable.*

The details of what is now called recursive unsolvability were conveyed to me in Cambridge, Massachusetts one winter day in 1936 by a stranger impertinently calling himself Juan Luis Borges, who, although quite obviously English judging by his worsted suit and very pale skin, bore no relation, or so he claimed, to the logician Alan Turing, who at the time was said to reside in Princeton, New Jersey. Nonetheless, I conceived the impression that Turing and Borges were somehow connected, the more so after it was revealed many years later that Borges's appearance in Cambridge coincided with Turing's unexplained disappearance from Princeton during the course of a long distance race.

"Are there tasks," Borges asked, as he lit and then extinguished one unfiltered Players after another, "that lie beyond the province of an algorithm?"

This is, to be sure, I observed, the same question that had antecedently been raised by the Cathar heretic Rues de Cervantes in 1257, who had sought to determine whether there might be limitations on divine competence; but de Cervantes, while he had conceived the question, was unable to provide the answer, inasmuch as he had perished in the inquisition launched by Innocent III.

"So it is often thought," Borges responded. "It is nonetheless the case that before his death de Cervantes was able to convey his conclusions to certain of his followers in Toulouse, who took it upon themselves to record his essential argument in Occitan. A summary of an Arabic translation of the original argument appears in a note written by Gottfried Leibniz to the French Jesuit and cryptographer, Jean-Luc Brice. In 1906, the contents of a steamer trunk belonging to one Ana Shpatalkova, a well-known Moscow ballerina and *demimondaine,* disclosed a Russian version of de Cervantes's original argument. Thereafter versions of the argument were known to circulate in the commons room of King's College in Cambridge, England."

I thought to comment on this strange narrative, but Borges waved his elegant but nicotine-stained finger in the air as if to suggest that comment would be superfluous.

"Whatever de Cervantes's original concerns," he said, "the

program conceived with the intention of bringing all of mathematics under algorithmic control has now reached the paradoxical conclusion that certain problems are insoluble by algorithmic means.

"The proof of this puzzling but profound result," he went on to say, "has an unexpected but welcome simplicity."

"The *proof*," I murmured. Borges waved his hand again, this time indicating a sense of urgency or perhaps impatience that he had so far failed to express. "The argument was first conveyed to the future by Rues de Cervantes. The proof is more recent."

I said nothing, but continued to listen.

"The imaginary machine devised by the logician Alan Turing," Borges said, "is devised to calculate functions, and it is only functions $f(x) = y$ from one natural number to another natural number that are under consideration. The natural numbers, Leopold Kronecker observed in Berlin, are gifts of God, everything else the work of man.

"The Turing machines taking one number to another number are now arranged in a Master List: there is the first M_1, and the second M_2, and the third M_3, and so on to the four hundredth and sixty fourth M_{464}, and then to the numerous points that like stars lie twinkling beyond. Each machine specifies an associated function *f*, the machine M_{23} acting to take a number *x*, for example, and then square it, while the machine M_{143} serves to add one to any number it contemplates."

I distinctly recall on hearing these words in Cambridge the extent to which they prefigured a story I would later write.

"It is here," Borges continued, "that a new numerical function *g* is introduced. It is a function that de Cervantes suggested would later be described as *inscrutable*. Like every other function, *g* takes numbers to numbers. Its secret nature, for like all aspects of the divine numerical functions have a secret nature, is determined by a simple rule, although one concealed for many years from all but a small number of logicians. Strangely enough, the rule is itself organized as an algorithm."

And here Borges wrote the algorithm in the air with his fingertips:

> *In order to determine* the value of g at the number x,
> *first* consult the Master List of functions $f_1, f_2, f_3, \ldots, f_n$.
> *Next* consider the function f on the list that corresponds to x.
> *Then* compute the value of g at x by
> *Setting* $g(x) = f_x(x) + 1$.

"Plainly," he said by way of explanation, "if $x = 10$, then $g(10) = f_{10}(10) + 1$. It may happen, of course, that at x, $f(x)$ is not defined at all, as occurs when $f(x) = x - 1$, a formula descending into incoherence when $x = 0$."

If that is the case, *then* the value of g is 0.

"The function g," he then observed, "is defined for every number; but curiously enough, inasmuch as it has been very explicitly defined, no scrutiny of the Master List of computable functions suffices to reveal the presence of g on the list itself."

It had begun lightly to snow.

Closing the collar of his vicuña coat around his throat, Borges stroked his eyebrows so that they lay flat against his forehead.

"The proof proceeds by contradiction, the strategy, it will be recalled, favored by Thomas of Aquinas in Book Two of his *Summa Contra Gentiles*.

"For suppose that g *were* on the Master List. It must thus be identical to some computable function f – the 17th, say. For every number x, it follows again that $g(x) = f_{17}(x)$.

"And since g has been defined for *every* number, it follows in particular that $g(17) = f_{17}(17)$.

"But from the algorithm defining g, it follows as well that $g(17) = f_{17}(17) + 1$. If $f_{17}(17)$ is not defined, then $g(17) = 0$. Whether equal to $f_{17}(17) + 1$ or to 0, $g(17)$ is *not* equal to $f_{17}(17)$.

"A contradiction has been reached."

And thereupon Borges paused to allow me time to reflect upon his words.

"You see," he finally said, "the only conclusion consistent with these facts is that g is *not* on the Master List at all. Thus g is not computable and since the Master List contains all of the Turing machines as well, it is in particular not computable by *any* Turing machine. It is for this reason, as Rues de Cervantes understood, that like the Deity itself, g must be considered inscrutable."

"And it was for this reason," I said, as snow began to fall more thickly, "that de Cervantes was punished?"

"It was reason enough," Borges said enigmatically; he then walked briskly from the park bench on which he had been sitting and quickly disappeared into the gathering shadows.

■ ■ ■ ■

It remains for me to act as Borges's amanuensis, pointing out pedantically that the argument he has recounted is simply a variant of Cantor's famous diagonal argument. What gives to the argument its maddeningly enigmatic air is just the fact that inscrutability has been established by what would appear to be a perfectly ordinary algorithm—the very one defining g, in fact. This is a puzzling circumstance. I am content to leave it at that, leaving it to Borges to unravel the consequences of an argument he never knew but apparently understood.

Whatever their nature, the non-computable functions serve to indicate certain intellectual limits, places beyond which human artifacts—and what is an algorithm if not a human artifact?—lose their power to coordinate and control events. The existence of such limits has by now become a familiar fact. The laws of physics mark the edge of the intellectual world; beyond them, there is nothing. The powerful and disturbing analysis provided by mathematical logic in the 1930s simply introduces us to limits of a further kind. For reasons we cannot quite fathom, they are transparent, leaving us the tantalizing prospect of being able to peep through them without being able to pass through them. Gödel demonstrated not just that arithmetic was incomplete, but that the sentence affirming its incompleteness was *true*. The existence of inscrutable functions demonstrates that the method intended to ratify the results of mathematical reasoning goes unratified, but this by means of the very method itself.

There are in these results intimations of the arbitrary sharply at odds with the history and very nature of science itself; but if the results achieved by logicians in the mid-1930s are negative, they are also liberating in their capacity to release by means of the pliers of paradox the grip of a certain system of illusions, and since what has been achieved has been *achieved*, who is to say that the light does not outweigh the darkness?

CHAPTER 12

Time Against Time

The process by which Turing's imaginary machine found itself embodied in a number of increasingly effective computers is a story that has itself been told a dozen times, occasionally by the men who did the work and occasionally by men who observed the men who did the work, and wrote of what they saw. It is one of the oddities of history that this story is itself rather dull. The first computers were slow and clumsy, vacuum tubes doing work that would later be done by transistors; stored programs made their appearance late in the 1940s, and then transistors and integrated chips a decade later. Over the next forty years computers got faster and they got better, and contrary to every reasonable expectation, they got more reliable, as well; and although modern computers crash, they almost never fail.

Within the United States, the architecture of the digital computer was from the first influenced by the Hungarian mathematician John von Neumann, who, dressed always in three-piece suits like an investment banker, was one of the remarkable intellects of the twentieth century, his brain, by common consent, an organ of astonishing versatility, capable of grasping and solving any conceivable problem with almost supernatural speed and accuracy. Von Neumann lacked the supreme genius of Gödel or the imaginative and radical simplicity of Turing or even the stolid monumentality of Alonzo Church, but he made up in speed and breadth and mathematical technique what he lacked in depth, and he defended his vision of digital computation with a combination of intellectual force and diplomatic skill that seemed quite compelling to both senior American bureaucrats and army generals, the first of whom he awed and the second whom he flattered.

But all this is something we already know; what is interesting, and what haunts the imagination is something else at work, some re-registration of the nature of experience that has been brought about by the advent of the algorithm, and so by the contrivances of art. For it is the algorithm that rules the world itself, insinuating itself into every device and every discussion or diagnosis, offering advice and making decisions, maintaining its presence in every transaction, carrying out dizzying computations, arming and then aiming cruise missiles, bringing the dinosaurs back to life on film or showing John F. Kennedy shaking hands with Tom Hanks, predicting the weather and the global climate, and like blind Tiresias, foretelling the extinction of the universe either in a cosmic crunch or in one of those flaccid affairs in which after a very long time things just peter out.

WORLDS OF OUR MAKING

■ ■ ■ ■ Confronting what seemed to be the plain fact that things fall apart, the nineteenth-century physicist Ludwig Boltzmann invented thermodynamics to explain what he thought he saw; and, then, confusing Vienna for the universe, he despaired at the predictions of his own theory, taking his own life well before the universe-as-a whole had time to gutter itself out in an unpleasant heat death. I don't suppose that Boltzmann ever thought his suicide a self-fulfilling prophecy, his personal encounter with the second law of thermodynamics as much an artifact of his own theory as anything else. There are, after all, countlessly many worlds in which things organize themselves for the better and not for the worse. A child is born; a book scrambles itself together from nothing more than symbols, ink, and the imagination; and if the entropy of the physical world is running down, it must at some time have been running up. There are worlds and there are worlds, and physics describes one of them, but *only* one of them.

Works of the imagination—stories, poems, pictures, music—describe the others; and we turn to them for relief; but the algorithm, that, too, is a work of the imagination, one of matchless powers, and as it comes shyly to be deployed in the serious sciences, it pries open a new place and bends time itself in new ways.

Is it logically possible, Vladimir Nabokov once asked, to tunnel backward in time and shake hands cordially with one's great grandfather? I have no idea, of course. The very thought of appearing in the early 1850s and clapping the old boy on the back is intellectually intolerable to me; yet in my dealings with Heloïse I was haunted always by the feeling that I could perceive her past in my future, a circumstance that seemed inevitably to involve a delicate double vision in which the real Heloïse, standing before me and stamping her foot with impatience, would be juxtaposed against a Heloïse younger than herself, the two women occupying companionable spaces in a universe in which time had somehow curved backward like a bright bow.

I had sailed from New York that autumn on the flagship of the French fleet, an elegant liner with three red smokestacks (they had figured, those smokestacks, in a certain cigarette advertisement) and a general air, despite its elaborate refurbishment the year before, of having seen too much and sailed too long. (Sometime later, the thing sank ignominiously at berth in New York, its interior gutted by a frantic and mysterious fire.) The ship's horns honked mournfully, a deep, bullfroggish sound—*farewell—forlorn*; the engines, which had until then been idling, commenced a low, deep, vibrant roar; the metal plates upon the deck shuddered; and after a few moments, a school of amiable, bright red tugboats maneuvered the monster into midchannel. A curtain of wet, gray clouds fell between the ship and the shore.

I shared my cabin with a taciturn Frenchman and a boisterous, manly, athletic Christian—a well-known shirt model, in fact—who greeted me with a booming "Praise-the-Lord" and an invitation, which I promptly declined, to join him in prayer. The fourth bunk in our room stood unoccupied, the bed representing somehow a reproach and serving after a time sullenly to collect our toilet articles and used towels and copies of a shipboard newspaper optimistically entitled *Bonjour mes amis!* A sign in the bathroom composed in three languages (French, English, Italian) warned passengers not to drink from the faucets. After my first shower, the water turning instantly from a precisely composed blend of hot and cold to North Atlantic frigid, I was advised solemnly by the Frenchman (his hands waving like pantomimic serpents as he delivered these asseverations) that I had been bathing in the ship's wastes, urine mostly; despite his beady, close-together eyes and sloping forehead, he spoke with the confidence of someone privy to a great many secrets.

A squall came and went. The air acquired a dense, gunmetal gray sheen, the sky simply merging moistly into the sea.

The ship's stabilizers, a steward gravely announced, had been extended. There were two seatings for dinner. I ate lavishly, confusing the first course for the whole meal, and then was lavishly sick on the middeck, coughing out my *sauce bernaise* in the direction, I thought, of Bermuda.

Later in the evening, I wandered into the ship's lounge. There at the end of the bar, looking cool and collected, rose-lipped and radiant, was Heloïse. The faint, fragrant flush of recognition that I felt had nothing to do merely with memory: I am not talking of the garden by the terrace, or the lilacs in flower, or a honey-heavy voice calling *"Rachel"* with vast poignancy and a sense of irretrievable loss.

She was perhaps a feline forty. She had sedate clear gray eyes and heavy copper-blonde hair swept up in a chignon and pinned with a diamond clasp; she was dressed in a very elegant gray sheath of a suit, which rode over a lime-green silk blouse, open at the throat; and she was talking to the ship's barman, who stood by the edge of the bar, his hip braced against the ship's roll, polishing a glass over and over again, his heavy head nodding absently from time to time. Her hand emerged to rest lightly on his forearm as if to detain him. She had that forwardness of gesture that stands midway between pushiness and poise.

She caught my eye and smiled.

I crossed from my end of the bar to hers and asked her to dance.

I had never seen her before. It was all familiar to me. Her size and smile, the easy indolent way she slid into my arms, the rolling ripple in the small of her back. The line of her jaw, I observed with a feeling of mute dismay, was beginning to sag. She wore a perfume I have since come to recognize as Chloé.

I slid my nose into the hot hollow of her neck; she moved her head backward in a tight tense little gesture; I could smell in the lines of her cheek the powder she had dusted across her face; and by means of a dancing maneuver in which I initiated but did not complete a turn, I managed to compress her body against my own.

"Whatever are you doing?" she asked with real astonishment, the Heloïse of Nabokov's story wondering just who that stranger was walking forward along time's axis even as he seemed to himself to be walking backward.

■ ■ ■ ■

Not all worlds are like that, of course. In the nineteenth century, thermodynamics for the first time brought an intellectual order to what is in a certain sense the most obvious feature of certain experiences. In very many purely local contexts, things seem to fall apart. The dealer's cards scramble themselves after shuffling; coffee turns lukewarm, and so, too, love; and the delicate crystal figurine fails forever to reassemble itself after being dropped. Disorder proceeds relentlessly, especially in rooms and lives, and physicists, no less than novelists, having noticed this world, are eager to give it an explanation. The story is by no means a simple one, but it has a simple denouement, as when several complex narrative lines resolve themselves in the passage of a well-made story. The arrow of time has its explanation in terms of probability, things falling apart for no better reason than they are likely to fall apart.

The proverbial and certainly the textbook example concerns the behavior of an ideal gas, one whose molecules comport themselves within a fixed and bounded region of space—a box, say. The molecules are themselves independent: what they do, they do on their own, banging about the place more or less randomly.

Imagine such a gas composed of just ten molecules, and imagine as well that the box in which they have been placed has been separated into two halves by a partition. At any given time, one and only one molecule may travel from the right side of the box to the left, or the reverse. Any particular distribution of molecules to the two sides of the partition is known as a *configuration*.

Time is now engaged; the molecules shuffle about the box, changing their configuration randomly, and therewith arises the question of the disposition of those configurations after time has gone on and gone by. Intuition has a surprisingly strong voice in the analysis, mine at any rate. If all ten molecules are on the right (or the left) of the partition, intuition requires—it *demands*—that after a decent period of time, the system as a whole should tend toward an equilibrium in which roughly half the molecules are on one side and half on the other. What intuition fails to supply is an accounting for itself.

Suppose that as time commences, all ten molecules *are* located within the right partition. Only one possibility prevails. A single molecule may migrate westward, crossing the partition and ending on the left side of the box. If at time zero, there are ten molecules on the right side of the partition, at the next instant of time, the probability is ten in ten that there will be only nine, the same inexorable probability establishing that there must be just one molecule on the left side of the partition.

The next instant, things change again. Another molecule may migrate westward, but then again, the first molecule may wander eastward, emptying the left partition. With respect to any particular right-sided molecule, the odds in favor of a move westward lie at nine in ten with respect to the single left-sided molecule, the odds in favor of the reverse are one in ten.

This exercise can be continued for a few more steps, with results that are anything but surprising; and if the calculation is prolonged, the average number of molecules on each side of the partition converges to a neighborhood of five.

But here is the interesting point, the very place where systems of description fail to cohere into a wave. The computations very soon grow intractable, and if the number of molecules is large to begin with, they become impossible, those calculations, even for a powerful mainframe computer running until the end of time.

It is for this reason that physicists turn to simulation in order accurately to gauge the thermodynamic behavior of molecules wandering vagrantly in an otherwise empty box. An easy method of simulation involves what is known in the trade as the Monte Carlo method. Using the method, their shirt cuffs discreetly pulled back, physicists employ a two-faced technique. First they generate a random sample of possible molecular moves; and then they assume that the sample is somehow or for some reason representative of the whole ensemble. An algorithm is introduced to handle the details, and like everyone ostensibly handling only the details, the algorithm is soon handling the main part of the drama as well. The algorithm works by means of a single controlling assumption: The probability that any given particle will move either to the left or to the right of the partition lies at n/N, where n represents the number of particles on either side of the partition, and N is the total number of particles altogether.

Herewith the algorithm in outline:

---

1. PULL a random number R from a hat (or any other device generating random numbers);
2. COMPARE R to n/N;
3. MOVE a particle from the left to the right if R is less than n/N; or do the reverse if R is greater; or do nothing if R equals n/N.

And herewith in more detail:

```
PROGRAM box
RANDOMIZE
CALL initial(N,tmax)                    ! input data
CALL move(N,tmax)                       ! move particles through hole
END

SUB initial(N,tmax)
  INPUT prompt "number of particles = ": N        ! try N = 1000
  LET tmax = 10*N
  SET window -0.1*tmax,1.1*tmax,-0.1*N,1.1*N
  BOX LINES 0,tmax,0,N
  PLOT 0,N;
END SUB

SUB move(N,tmax)
  LET nl = N                            ! initially all particles on left side
  FOR itime = 1 to tmax
    LET prob = nl/N
    ! generate random number and move particle
    IF rnd <= prob then
      LET nl = nl - 1
    ELSE
      LET nl = nl + 1
    END IF
    PLOT itime,nl;
  NEXT itime
END SUB*
```

*Program box algorithm. Harvey Gould and Jan Tobochnik, *An Introduction to Computer Simulation Methods; Applications to Physical Systems*, vol. 2 (Reading, Mass.: Addison-Wesley, 1988): 487.

The algorithm now engages itself, playing over an algorithmic universe, and once again reaching results that are in alignment with intuition. Reasoning that is primitive but persuasive has given evidence that moving in accord with certain laws of probability, imaginary particles within an imaginary box separated by an imaginary partition will move toward equilibrium when given life within an algorithm.

■ ■ ■ ■ Or in a story. Some twelve years later, upon the completion of a jump cut in which earlier and later portions of my own life were cleverly spliced, I crossed West 87th Street in Manhattan on the morning of a cold, clear, cutting day late in the fall. I had been married and was again single; and I needed to purchase a pack of cigarettes. Walking toward me with the brisk, determined strides of a much younger woman was Heloïse. She had gained perhaps a half inch in height; her face had tightened; she had lost that very faint, very fine droopiness along the margins of her jaw; she wore no lipstick, only lip gloss; and she was dressed in jeans and a red flannel shirt, open at the throat. I followed her, of course, down Broadway to 86th Street, and then over to West End Avenue; she turned into the very building in which I lived.

She was, she told me two weeks later, a singer, a lyric soprano, and a student of some fat fraud then in the stage-managed sunset of her own career. She would be auditioning for the Metropolitan Opera in June; she was prone to sore throats and loved stuffed pandas. I said something sincere and sympathetic and wished only to press my lips to the trailing edge of that delicate bone that rode from her shoulder to the vertex of her throat; but when I leaned over, she leaned back; and it took me almost two more weeks to maneuver her from the twelfth floor of the building in which we both lived to my bedroom on the tenth.

Later that spring we drove from New York to the northern

Catskill Mountains. We traveled in a rented car through Oxblood, Katterskill, and Phoenicia, ancient, crumbling, Dutch towns in their gloomy cast and character, located at the irresolute ends of old horse trails, Phoenicia, in particular, consisting of a single set of wooden stores, a post office, the rotting hulk of a blacksmith's stable, an IGO market beside a gasoline station (one pump, forlorn), the road itself passing over an angry little mountain brook by means of a concrete WPA bridge. Two or three miles past Phoenicia, as the ancient highway rose and clattered and ascended its way up the side of the mountain, the air fragrant and wanton, we came upon a dilapidated cluster of small Komfy Kabins, all wood, pitch, and a faint, elusive, sewery smell, the drainage field, no doubt; they straggled, those cabins, along a limestone field.

Standing in front of an enormous moose head (the thing fixing everyone with a single glaring and accusing plastic eye), the proprietor of those estates, a very rugged-looking character, with a narrow, almost hairless head, small eyes, rather a beaky nose, and leathery cheeks, ran a clubbed index finger over his tensed lips and said, "Sure I could let you have one of them cabins for a few days, young fellah." If he had any interest in the young woman who sat in the front seat of the car in an agony of embarrassment, and whom I proposed to smuggle into the cabin as soon as I managed to complete this odious ritual, he did not let on; for the rest of our stay he was the very model of rural gentility. As soon as he received payment for the cabin (thirty dollars, I think, a pink pittance), he disappeared into the back of his house, which contained the office as an afterthought—and from which a deep, greasy smell emerged—and did not reappear ever again.

"I am *so* nervous," said Heloïse, when we finally entered the cabin together. The thing was even dingier on the inside than it appeared from the little lawn. There was no electricity, but our host had tendered me a kerosene lantern, and there was no toilet, either, a circumstance that Hank or Hiram or Horace suggested we confront by a fluid wave of his outstretched hand pointing to the woods beyond. The ancient mattress on the iron bed had obviously been the haunt of a family of enterprising mice. Their litter was everywhere, and so we repaired to the car, fetched our sleeping bags, and fashioned a makeshift mattress on the creaking springs themselves.

We stood together in the dark little cabin, the musty woodsy smells around us. Heloïse was somber and sensitive, and perfectly poised.

■ ■ ■ ■

Within thermodynamics, equilibrium denotes a balance of sorts, forces tugging left just matched by forces tugging right, and thus forcing a system of particles, or a pair of characters, into a moment of equipoise. The great physicists demand more from their theories, some deep current of metaphysical respectability. There is no gainsaying the fact that things, but not children, tend to fall apart; and this fact, although explained in part by cooked-up cases, seems to engage itself with the deeply mysterious and much larger fact that time has a direction in the universe, moving forward if it moves at all. With the dramatic entrance of these facts, thermodynamics reaches for the concept of entropy.

The same setup as before, sustained by the same discordant strangers—probability, the algorithm, the imagination. Suppose that there are only two molecules in the box, these labeled **A** and **B**. Time is on hold, allowed to move forward only when the physicist releases its restraining ratchet. The molecules are for all practical purposes indistinguishable, differing only in their position. There are as a result four and only four configurations accessible to the system in the box. **A** and **B** may be together on the right side of the partition or on the left, or they may divide themselves on either side. These possibilities are listed in a notation that explains itself: **AB**; **A—B**; **B—A**; **BA**.

Now if **A** and **B** are moving randomly, and so have a one-in-two probability of being on either side of the partition, each of these four possibilities has precisely the same probability, namely one in four.

And yet, as Boltzmann noticed, there is a sense in which the world represented by those four molecules, arranged in four possible configurations within the confines of a single divided box—the entire universe imaginary—gives just the most subtle hint of descriptive impropriety; for of these four configurations, two are virtually indistinguishable because **A** and **B** are virtually identical.

Looking at the system from a *macroscopic* point of view, one in which temperature or pressure or energy is measured, the physicist cannot tell whether **A—B** or **B—A** prevails. Following Boltzmann, he thus comes to the radical and unexpected conclusion that while the system has four microstates (**AB**; **A—B**; **B—A**; **BA**), it has only three macrostates (**AB**, **BA**, and **A—B** *or* **B—A**). This is a perception of exquisite fineness.

The way in which configurations and macrostates vary can be expressed by a simple mathematical observation, one entirely combinatorial in nature. If there are N molecules to begin with, where N is 10 or 10,000, and n represents the number of those molecules to the right of the partition and n^* the number to the left, then the number of configurations corresponding to each macrostate is given by the binomial distribution

$$\frac{N!}{(n! \times n^*!)}$$

Memory, which has so far supplied this little lecture now supplies its antecedent groan. Like so many formulas in mathematics or the sciences in general, this one requires attention before it begins to vibrate in all the colors of the spectrum.

But vibrate it does. Thus in my own little cooked-up case, there were two molecules in all, **A** and **B**. The factorial of a number (denoted $N!$) is the product of that number and each of the numbers less than it, with 0! defined as 1. $N!$ is thus simply 2. The factorial of $n!$ must be 1. If n and n^* are both 1, $N!/(n! \times n^*!) = 2$, and there are *two* configurations compatible with the one democratic macrostate. If **A** and **B** are segregated together on either the left or the right side of the partition, then either $n = 2$ or $n^* = 2$, and in either case, $N!/(n! \times n^*!) = 1$. There is only *one* configuration compatible with each of the other macrostates.

There now follows yet another imaginative act. That box has thus far been all bouncing balls, confining walls, and random energy. What else could there be? And yet with the work of other subtle and ingenious novelists at his disposal, Boltzmann urged otherwise, investing that closed container of the imagination with an entirely abstract *something*, one capable of growing or subsiding, and one capable of being expressed as a quantitative property, something belonging to the broad and noble class of properties that are measurable.

Such is the entropy of the system.

The relationship between the entropy of a system and its configurations, Boltzmann expressed by means of the formalism of a mathematical law. For any given macroscopic state n of a system of N interacting particles, the entropy of n is proportional to the logarithm of Ω_n, where Ω_n is simply the full set of configurations compatible with n. (Boltzmann chose the logarithm to slow descriptions of the system, since the logarithmic function grows slowly at first.)

In symbols:

$$\textbf{Entropy of } n = k \log \Omega_n.$$

Here k is Boltzmann's constant, a fixed number intended to simplify internal mathematical manipulations.

The law suggested to Boltzmann a far-reaching generalization of a limited physical relationship. For one thing, entropy is a quantity that can be expected to increase remorselessly. Disorder proceeds apace if only because disorder admits of multiple realizations. That far-reaching generalization now follows: Although the examples that went into the construction of thermodynamics were drawn from the theory of perfect gases, the underlying—the crucial—idea was expressed by the isolation of the system from alien influences. Turning his thoughts upward and outward, Boltzmann realized that the universe itself is precisely such an isolated system; after all, if the universe is everything there is, what possible alien influences could affect its behavior? If entropy increases in closed systems, it must increase everywhere else as well, the everywhere else taking in the glowing sun, the solar system, the Milky Way, the galaxies beyond, and finally the whole far-flung structure of space and time. These too must tend inexorably toward their most likely state, one in which entropy is at a maximum, the universe at the end of time vacant, motionless, uninteresting, flaccid, pale, and flat.

▪ ▪ ▪ ▪ That night we lay in one another's arms, awake and then asleep, wondering from time to time just what was making those strange and alien noises in the woods and fields, and finally determining that it was an owl or another night bird or an insect with enormous mandibles or a raccoon or absolutely none of these, something sinister instead, a tiger, perhaps, or a snow leopard, with evil, gleaming yellow Stalin eyes.

When we awoke again the sun was high in the sky, and an immensely irritable fly was circling the soiled and heaped-together sleeping bags, making, whenever he thought we weren't looking, for the cupcake crumbs left over from the night before.

That afternoon, the weather turned humid; by evening great thunderheads had built up over the mountains; at night, it rained steadily; we were both constipated, unwashed, and when we returned to New York, Heloïse held my head in her long-fingered hands and kissed my eyelids with abstract poignancy.

■ ■ ■ ■

And yet, the relationship that Boltzmann proposed between the number of configurations latent within a closed system and its macrostates is metaphysical in the sense that for any realistic system, there is simply no way in which physically to count the relevant configurations. Before the advent of the algorithm, thermodynamics, like so much of classical physics, exhibited remarkable explanatory powers, but no concomitant scheme of detailed verification. Things fall apart. We can see that. Disorder gains over order. We can see that, too; but entropy itself, however much it may have acquired a numerical identity, remains inaccessible if only because in reasonably large systems, it denotes a quantity that is growing in ways we can sense but not see in any detail.

Whatever the power of any computer—now or in the future—it is not possible to deal with entropy directly: the numbers of microstates in any ensemble of particles is simply too large. Only the real world is capable of faithfully simulating itself, an odd and unsettling fact, and one that should be more widely understood. Nonetheless, the advent of the algorithm has made it possible to explore thermodynamics *in*directly, by means of a deliberate but contrived inspection of the microstatic behavior of a system of interacting particles.

The problem of simulating a system that one cannot see is by no means trivial—in physics or in art. An algorithm, or a story, must create a world. It is just such a world that the physicist Shang-Keng Ma created, his idea based on the old notion of recurrence. A system of molecules shuffling forward in time enters now one configuration and now another; but no matter the size of the system, there are—there *can* be—only a finite number of configurations. Recurrence is thus a feature of any finite dynamical system, the arrangement of molecules taking place in some closed container destined to reappear in the far, far future. This simple fact is the basis of Poincaré's famous recurrence theorem, the theorem saying roughly that sooner or later, whatever it was that was is destined to return, a prospect that leaves some of us with feelings of distinct unease.

It was Ma's simple but very ingenious idea to measure entropy as an observable property of a computer simulation by looking for coincidences between uncorrelated microstates. Digital time is allowed to advance; the search is made intermittently for coincidences to appear. The longer time proceeds without coincidences, the fewer the requisite microstates, and thus the lower the entropy of the system. The greater the number of coincidences, the greater the number of requisite microstates, and hence the greater the entropy.

Instead of dealing directly with entropy, and so with the real world, Ma defined a quantity he designated the coincidence rate, R_n, of an ensemble of simulated molecules. With the physicist poised to dip into and sample the stream of simulated time, R_n denotes the ratio between duplicated microstates and the total number of microstates that have been compared and contrasted. Thus if the physicist samples one hundred microstates and discovers that two are identical, R_n is just 2/100.

Using this ingenious idea, Ma replaced Boltzmann's law with his own, thus raising his own imagination to transcendent status:

$$\textbf{Simulated Entropy of } n = k \log \frac{1}{R_n}.$$

It is easy to mistake the relationship expressed by this law with the one that Boltzmann assumed was a law of nature; it is easy, as well, to assume that Ma is merely presenting a scheme for the solution of classical thermodynamic equations. But this is not so. Simulated entropy is a property of a simulated world. Within the confines of this world, Ma is in charge, his control over time and decay absolute. It is commonplace to call a world of this sort virtual, as if it were able only to catch a pale imitation of the real world's fires; but there is nothing virtual about any of it. In many respects, it is far more accessible than that other world, the one in which we live and die and which hides its particles in clouds of unknowing and covers even its most obvious processes in numbers too great to notice. *This* world is entirely accessible, its laws perfectly lucid, and the processes that they describe entirely subservient to the laws that describe them.

With certain modifications, the program box algorithm already cited will do the job of simulation. Ma's definition is contingent upon the specification of some very particular macrostate: such is n. In order to simulate this contingency, the number of imaginary particles to the left and the right of an imaginary partition is held constant. This brings about the required specificity in states. Thereafter, the system evolves by an *exchange* of particles, rather as if its dynamical behavior were being held at an auction.

The program that follows actually computes the entropy for a simulated system, with results, I may as well add, that are entirely in accord with the standard predictions of thermodynamics.

```
PROGRAM ENTROPY
DIM left(10),right(10),micro(0 to 2000)
RANDOMIZE
! input parameters and choose initial configuration of particles
CALL initial(nl,nr,left,right,micro,nexch)
CALL exchange(nl,nr,nexch,left,right,micro)          ! exchange particles
CALL output(nexch,micro)        ! compute coincidence rate and entropy
END

SUB initial(nl,nr,left(),right(),micro(),nexch)
   ! fix macrostate
```

```
  INPUT prompt "total number of particles = ": N
  INPUT prompt "number of particles on the left = ": nl
  LET nr = N - nl                ! number of particles on the right
  LET micro(0) = 0
  FOR il = 1 to nl
     LET left(il) = il           ! list of particle numbers on left side
     LET micro(0) = micro(0) + 2^il                 ! initial microstate
  NEXT il
  FOR ir = 1 to nr
     LET right(ir) = ir + nl     ! list of particle numbers on right side
  NEXT ir
  INPUT prompt "number of exchanges = ": nexch
END SUB


SUB exchange(nl,nr,nexch,left(),right(),micro()
  ! exchange particle number on left corresponding to lindex
  ! with particle on right corresponding to rindex
  FOR iexch = 1 to nexch
     ! randomly choose array indexes
     LET lindex = int(rnd*nl + 1)                  ! left index of array
     LET rindex = int(rnd*nr + 1)                  ! right index of array
     LET left_particle = left(lindex)
     LET right_particle = right(rindex)
     LET left(lindex) = right_particle   ! new particle number in left array
     LET right(rindex) = left_particle   ! new particle number in right
                                           array
     LET micro(iexch) = micro(iexch - 1) + 2^right_particle
     LET micro(iexch) = micro(iexch) - 2^left_particle    ! new microstate
  NEXT iexch
END SUB

SUB output(nexch,micro())
  ! compute coincidence rate and entropy
```

```
   LET ncomparisons = nexch*(nexch - 1)/2           ! total number of
                                                    comparisons
   ! compare microstates
   FOR iexch = 1 to nexch - 1
      FOR jexch = iexch + 1 to nexch
         IF micro(iexch) = micro(jexch) then
            LET ncoincidences = ncoincidences + 1   ! number of
                                                    coincidences
         END IF
      NEXT jexch
   NEXT iexch
   LET rate = ncoincidences/ncomparisons            ! coincidence rate
   IF rate > 0 then LET S = log(1/rate)
   PRINT "estimate of entropy = ",S
END SUB*
```

Ma's law is designed to shed light on the microscopic structure of a toy universe that in its turn is intended to be a surrogate for the microscopic structure of Boltzmann's toy universe. But this curiously satisfying concordance between worlds has been purchased by means of assumptions that while they seem reasonable are nonetheless arbitrary. A cascade of mathematical theories has been introduced to deal with experience, and not all of the theories are cut from the same cloth. The experiences which are the subjects of our own very personal complaints, commiserations, or congratulations have been cut and trimmed and redescribed so that having appeared originally in one story they may emerge later in another. In order to get anything done at all, the physicist has been forced to rely on the algorithm, an artifact alien to any of the activities that he means to describe, with results that compromise, and compromise fatally, that in-

*Program entropy (computer language: True BASIC). Harvey Gould and Jan Tobochnik, *An Introduction to Computer Simulation Methods; Applications to Physical Systems*, vol. 2 (Reading, Mass.: Addison-Wesley, 1988): 493.

nocent sure sense that there is in any of this one waiting world, one system of description, one resolution of experience.

When physical processes are not so much specified as imagined, just who is the master and who the mastered?

■ ■ ■ ■ Twelve years after Heloïse's kiss, my eyelids still burning, I found myself in San Francisco. The woman for whose sake I had tediously crossed a continent disappeared two months after I arrived in California; a private detective named Asherfeld located her finally in an ashram in Oregon. I had been married again before *that,* and divorced, my former wives (three in all) forming a little society in which complaints about me would ricochet across the continent.

The department of mathematics at Berkeley had invited me to deliver an address; I needed to organize my thoughts: I had walked without knowing why from my apartment in Pacific Heights to the eastern edge of the Golden Gate Park. It was a calm, perfectly coordinated day. The sky was that thin-tin maddeningly elusive blue characteristic of California; the grass in high winter was green, vibrant, and an unusual procession of melodramatic cumulonimbus clouds, bewildered refugees, I imagine, from the Midwest, had piled upon the horizon. The ramble that runs through the park and that ends ultimately by an ocean beach, the abysmal haunt in that part of the city of adolescent drug dealers and aimless loping dogs depositing their wastes on the yellow sand, was closed to traffic. Just in front of the lacy white arboretum the roadway acquires an aneurysm to accommodate a variety of side streets bent ultimately on obliterating themselves by anastomosis. On weekends roller skaters gather here to flaunt themselves and practice. Someone had installed an enormous set of speakers by the side of the road; they were, those speakers, pumping a heavy-lidded, maniacally monotonous rhythm into the crisp air. Two rangy black youths described linked loops on the walkway, drifting from one side of the road to the other, skating backward all the time. A fiercely determined, middle-aged white man, all compact, bunched muscles and a look of intense, dentistlike concentration, was endeavoring to insinuate his rather gross, inevitably clumsy, movements into the fluid reaches of their graceful arabesques.

I stopped to take in the scene. The dentist skated over to the side of the road, his roller skates clattering, to deliver his dripping jersey to a camera-toting girlfriend; with his hands forming a director's square (thumbs pointing toward one another, index fingers raised), he indicated to her the superb panorama that he envisioned: the lucid green of the grass in the background and in the foreground he himself dancing like mad Nijinsky; the two blacks drifted from side to side like smoke.

What else? Let me see. There was a hot-dog stand beyond the roller-skating circle, and a secondary group of much younger exhibitionists doing tricks on squat bicycles, and a hugely pregnant young mother wheeling a baby carriage, a panting dog lumbering in her wake, and a man with a full beard carrying a green parrot on his shoulder.

A tall, angry-looking girl of perhaps eighteen, I now noticed, stood at the side of the roadway. She was dressed in iridescent hot-pink shorts; she was on roller skates and maneuvering with most of her torso simply to stay in one place; she had streaked orange hair and wore a good deal of flat pancake makeup. She was quite evidently indignant about something, for she stood in that rootless pose delivering herself of a perfect volley of remonstrations in a singsong foreign language.

The object of her indignation sat crumpled in a leggy heap on the curb. It was Heloïse, of course, perhaps three-fourths her sister's age (I am guessing about the relationship), but just as tall, judging things from the impression she gave, even while seated, of smooth lustrous length. She was dressed like her sister in pink satin shorts and a ribbed torso-hugging shirt; she wore a bracelet made of porcelain squares on her thin wrist and tiny, button earrings in her seashell ears; her legs were folded at her delicate antelope knees, monkey feet in white, laced, roller-skating boots, the boots themselves tucked underneath her thighs. Her lovely, limpid entirely puerile face was contorted by a spasm of unendurable grief. Her eyes were gray, the lower lashes smudged and sooty, her lips peeled-peach raw, as if she had been biting them, the delicate bluish skin of her tensed and flaring nostrils chapped.

And here is what she did (as time crawled and then stopped): she waved her hands before her face and folded them into her lap and lifted them into the air again and with a careless heart-heavy gesture wiped her smudged eyes and ran her fingertips through her chestnut hair and dropped her hands back into her lap, her shoulders flower-drooping.

She was whining something all the while in that singsong language of hers (Swedish, I think) and snuffling, the whining broken every now and then by a fresh set of sobs that shook her thin shoulders, and when her grief reached some melodramatic inner threshold she would draw in a great lungful of air, chin up, shoulders rising, and pivot her elbow in the hollow of her groin, holding her forearm to her chest, and slap the air (and by inference her sister beyond) fretfully with her long-fingered palm, the wrist limp.

That stolid sister, standing there unmoved by all that limpid radiance, continued to gabble and complain.

I stood there, heart-bruised myself; I wished only to fold this creature fourways into my arms (legs doubled underneath themselves, like a fawn, arms inward) and rock her gently and blow warm air into the silky hairs that lined the back of her neck and formed a musky golden halo.

"Is anything wrong?" I asked with magnificent irrelevance.

She looked up at me, confused for only a moment, her eyes wet, and smiled the very faintest most fragile and friable of smiles.

She rose to her feet, still hiccuping with indignation but calmer now, and after saying something suitably wounding to her sister commenced to skate eastward along the ramble, moving with the long uncoordinated strides of a distracted adolescent.

Nothing more. I stood on the roadway and watched her shrink in size until by means of some secret scheme or system she managed to transform herself from a leggy, very young woman on roller skates to a blur in the middle distance to a palpitating point (chestnut hair still flashing) on the farthest reaches of my line of sight.

■ ■ ■ ■

APPENDIX: WORLDS IN COLLISION

The world through which the physicist has wandered for over three hundred years without weariness is one made possible by the great mathematical apparatus of ordinary and partial differential equations; it goes without saying, of course, that physicists cannot solve most interesting differential equations and most interesting differential equations cannot analytically be solved.

Equations that cannot be analytically solved are analytically intractable; the information that they contain is shrouded by the cataract of our own mathematical incompetence. It comes as both a relief and as a warning to learn that intractable equations, while they cannot be solved may nonetheless be simulated. The point emerges most clearly in even a simple case, one in all the textbooks. It begins, as all scientific cases do, with a simple-sounding question: How does one describe a process of uniform growth (or decay)—the pattern exemplified, for example, by the behavior of a bacterial colony growing on a petri dish? Counting is one evident strategy, but beyond observing that there are more bacteria now than there were then, or fewer, as the case may be, this is a technique that seems scientifically unhappy because scientifically unwholesome. Too many bacteria, too little time.

What is wanted is a descriptive scheme that allows the mathematician predictively to compute the number of bacteria at some point in the future (or the past: time goes both ways) given *only* that scheme and the number of bacteria presently at hand. Note the tight and doubled restrictions: scheme and number must do all the work. When the problem is thus described, it acquires the familiar aspect of an initial-value problem, one represented mathematically by a very simple, very ordinary differential equation:

$$\frac{df(t)}{dt} = Af(t)$$

The expression $f(t)$ denotes a function and so a process, one mapping the time to the number or quantity of bacteria, so that if t is a particular time, $f(t)$ is a particular quantity, the genius of the notation lying in its ability to provide an endlessly extended scroll, one in which each moment in time is coordinated to some mass or weight of bacteria. Its mathematical nature as a function is for the moment unknown. The expression A is the name of a number, a so-called constant of proportionality, one indicating the proportion of bacteria engaged in reproduction; it is a number that remains fixed throughout this exercise so that no matter whether there are ten or ten thousand bacteria in that petri dish, only a fixed proportion of them are carrying on at any given time.

The expression $df(t)/dt$ is not quite a ratio, although it looks like one, but rather a single expression, the *derivative* of the function $f(t)$. It is intended within the mansion of the calculus to denote the rate of change taking place within f itself, as time slips forward continuously, and so is expressed as the *limit* of a sequence of ratios in which the quantity in question (in this case, the number of bugs) is divided by time as the elapsed time (indicated by Δt) proceeds toward 0:

$$\frac{df(t)}{dt} = \lim_{\Delta t \to 0} \frac{f(t_2) - f(t_1)}{\Delta t}$$

The concept of a limit, although defined precisely within the calculus by a complex of delicate constraints, refers nonetheless to an infinite process, the limit itself lying at the end of an infinite series of steps. The derivative of a function answers precisely at each and every moment of time the question of how rapidly or slowly some process is proceeding.

With this explanation on the table, the meaning of the differential equation now follows:

$$\frac{df(t)}{dt} = Af(t)$$

expresses the claim that the rate of change in $f(t)$, the number of bacteria at time t, is proportional at t to $f(t)$ itself.

And this makes sense. How fast a colony of bacteria will grow is contingent on the absolute number of bacteria on hand *and* the relative percentage of bacteria engaged in reproduction. How many? How much? The questions of trade; the questions of life.

Equations are by their very nature acts of specification in the dark; something answers to some condition. The condition or conditions are given; it is the something that needs to be found. Specification in the dark corresponds to the familiar and by no means mysterious process by which a sentence in which a pronoun figures—*He* smokes—acquires the stamp of specificity when the antecedent of the pronoun is dramatically or diffidently revealed—Winston Churchill, say, or a lapsed smoker sneaking an errant cigarette in the bathroom.

The differential equation describing uniform growth admits of a simple but utterly general solution by means of the exponential function

$$f(t) = ke^{At}.$$

The number e is an irrational number lying on the leeward side of the margin between 2 and 3 and playing, like π, a strange and essentially inscrutable role throughout all of mathematics; exponentiation takes e to a power of itself, in this case, one specified by the numbers A and t. The constant k has an interpretation as the problem's initial value, which consists of some enumeration of the number (or weight or mass) of bacteria.

And the indication that ke^{At} *is* the requisite solution to the original equation? Just that the calculus, once again, establishes that the derivative of ke^{At} is none other than $Af(t)$.

A differential equation has the uncanny power to penetrate the future (or the past, as it happens) and occupies its position in the physicist's pantheon for precisely this reason. The variable t takes time as its value, and as time scrolls backward or forward in the mathematician's imagination, ke^{At} provides a running account of growth or decay and so a running account of change.

This is in itself remarkable, the temporal control achieved by what are after all merely symbols quite unlike anything else in language or its lore or law; but when successful, specification in the dark achieves an analysis of experience that goes beyond any specific prediction to embrace a universe of possibilities loitering discreetly behind the scenes.

The fourteen symbols making up the equation

$$\frac{df(t)}{dt} = Af(t)$$

not only provide a tight quantitative control over the infinite extent of time: they *also* provide a wide-ranging and prophetic qualitative appraisal of the various ways in which time might play itself out. Things that increase in number, after all, may decrease in number as well, or they may remain unchanged. And these possibilities correspond to an imagined three-part separation of the past, the present, and the future. They are, those possibilities, arranged in virtue of the sign of A. Positive, A signifies growth; negative, A signifies decline; and if $A = 0$, nothing happens, and there is neither growth nor decline.

Growth, decline, and nothingness show themselves as graphs on a Cartesian coordinate system, each family of curves occupying some position in the large and handsome family of exponential functions (see figure 12.1).

The future has now been given a polychromatic cast, one even more remarkable than may be imagined in that the vivid hues of growth, decline, and nothingness serve to convey information not only about the future's shape but its stability as well. Stability returns the discussion to states, situations, or solutions that change their fundamental nature under small perturbations.

The differential equation governing uniform growth separates time into three phase portraits: two of these, growth and decay, are stable, small perturbations in the constant A governing that proverbial petri dish, serving only to change the rate but not the essential nature of the process underway. Stability prevails where there is change. Stasis is something else. If $A = 0$, the slightest deviation in its value changes, and changes utterly, the nature of the solutions. Something that has been snoozing sedately in the stream of time suddenly acquires the capacity either to grow or to decline. The polychromatic vision of the universe is an oblation within an oblation, like one of those elaborate, achingly beautiful Fabergé eggs in which an egg resides within an egg.

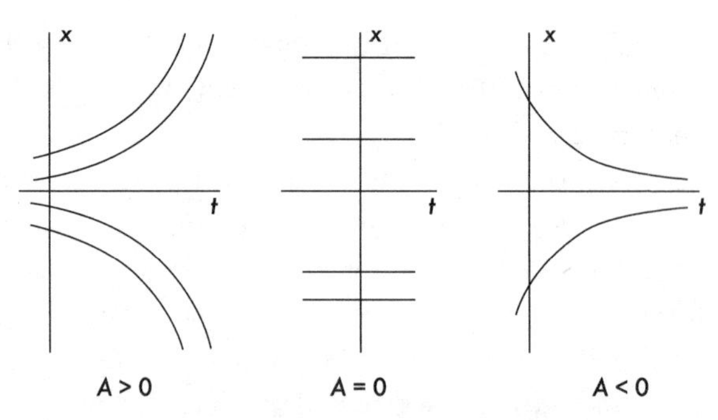

12.1. *Three phase portraits for* $\frac{df(t)}{dt} = Af(t)$.

Equations that cannot be analytically solved may nonetheless be simulated. Techniques for analytical simulation vary, but they are all constructed by means of an invocation of a case-specific algorithm by purely numerical, purely discrete, purely finite techniques of calculation. The most elementary of the techniques is, in fact, derived from the calculus itself, one reason the calculus is called the *calculus*.

The definite integral

$$F = \int_a^b f(x)dx$$

figures in the calculus both as the source of an analytic expression (F, of course, some function) and as a means to designate the area underneath a curve from points a to b; in this, its geometric incarnation, it has an identity as a limit of a certain sequence of sums:

$$\lim_{\Delta t \to 0} \sum_a^b \Delta t_i f(t_i).$$

The textbook examples show what is meant. The area underneath the curve is divided into rectangles, and as the rectangles become smaller and smaller, the approximation to the area underneath the curve becomes better and better. Better and better? Meaning better and better *as* the intervals become smaller and smaller, so that in the limit it becomes perfect and irrefragable, the limit representing not simply an approximation to the area but its real and absolute value (see figure 12.2).

There is a price, of course, to be paid for mathematical irrefragability, just as there is a price to be paid for everything and anything. One goes to the limit—strange the queer resonance of mathematical terms—by traversing an infinite number of steps. Some compromise with experience is required, some entry into an imaginary empyrean, for who among us can run a race that long?

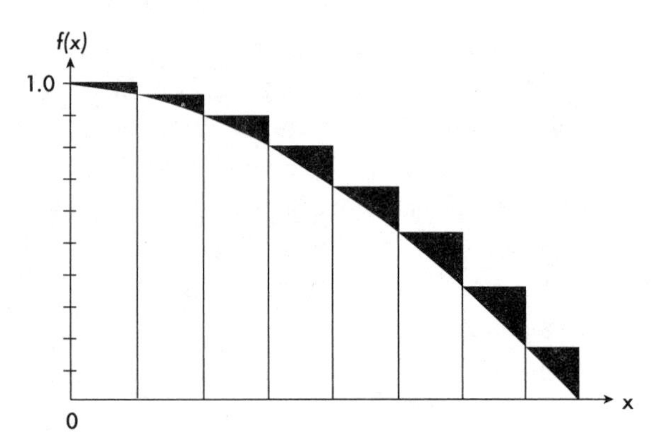

12.2. *Numerical integration.*

Nonetheless, the integral

$$F = \int_a^b f(x)dx$$

may numerically be calculated by means of a simple program, even when $f(t) = e^{-t}$. And as the reader must long ago have guessed, the formula for numerical integration is simply the formula for integration itself, shorn of the passage to the limit, so that

$$F = \sum_a^b \Delta t f(t).$$

The result is, of course, an *approximation* to the area underneath the curve, one made in full consciousness of the inevitable error that accompanies the approximation.

The approximation may be made better and better, but it may be made better and better in only a *finite* number of steps, and so remains within the haunted but human compass of purely human computational techniques.

The following simple algorithm performs precisely this task of approximation:

```
PROGRAM integ
! compute integral of f(x) from x = a to x = b
```

```
CALL initial(a,b,h,n)
CALL rectangle(a,b,h,n,area)
CALL output(area)
END

SUB initial(a,b,h,n)
   LET a = 0                          ! lower limit
   LET b = 0.5*pi                     ! upper limit
   INPUT prompt "number of intervals = ": n
   LET h = (b - a)/n                  ! mesh size
END SUB

SUB rectangle(a,b,h,n,area)
   DECLARE DEF f
   LET x = a
   FOR i = 0 to n - 1
      LET sum = sum + f(x)
      LET x = x + h
   NEXT i
   LET area = sum*h
END SUB

SUB output(area)
   PRINT using "####.#######": area
END SUB

DEF f(x) = cos(x)*
```

The same technique suffices to provide a numerical solution for ordinary differential equations that lack an analytic solution,

*Algorithm for numerical integration. Harvey Gould and Jan Tobochnik, *An Introduction to Computer Simulation Methods; Applications to Physical Systems*, vol. 2 (Reading, Mass.: Addison-Wesley, 1988): 321.

and so cannot be made to yield their polychromatic luster. Suppose that

$$\frac{df(t)}{dt} = g(t, y)$$

is precisely such an equation, an affirmation of change in which nothing accessible answers to $f(t)$—(the y, which stands for $f(t)$, within $g(t, y)$, by the way, making a more formal appearance on the equation's right side as a variable).

The search for an analytic answer having proved unavailable—I mean such is the assumption I am making—there remains the technique of numerical integration. There are many such techniques in the literature, but the key to all of them, of course, is approximation. Within the calculus, the derivative answers to the slope of a tangent line. The requisite approximation is constructed (or concocted) with this in mind:

$$\frac{df(t)}{dt} = g(t, y) = \frac{f(t_i + \Delta t) - f(t_i)}{\Delta t} \approx g(t_i, f(t_i)) + O(\Delta t)$$

Impressions to the contrary notwithstanding, there is nothing daunting about the details of this expression. Reading from left to right, the equation says that three things are *identical*. But this we already know. I have said it before and you have it on my say-so. What comes next is the symbol "$\approx$," which indicates approximation. The derivative of the function $f(t)$, the expression affirms, may be approximated by the value of $g(t_i, f(t_i))$ and an error term, which is designated by $O(\Delta t)$.

But by elementary algebra, this is equivalent to saying that

$$f(t_i + \Delta t) \approx f(t_i) + g(t_i, f(t_i)) \times \Delta t,$$

the error remaining what it was and where it was.

From a mathematical point of view, the original differential equation, contingent as it was upon the concept of a limit, has been replaced by a difference equation, one in which the derivative is approximated by a difference quotient involving no limits whatsoever.

The great virtue of the procedure is that it lends itself to implementation by algorithm, with each new solution point of the difference equation approximated by the tangent value at the point just given. Such is the Euler algorithm or method. I present it in what follows in an easily understood computer pidgen:

```
BEGIN Euler
INPUT x0, y0, xf, h
x : = x0
y : = y0
WHILE (x < xf ) DO
   y : = y + h*f(x,y)
   x : = x + h
   OUTPUT x, y
ENDDO
END Euler*
```

The Euler algorithm is by no means the sharpest or most efficient tool available; but in simple cases it works, and it is the simple cases that throw the sharpest but the harshest light. To say that the algorithm works is to say that within a finite period of time, and acting between two geometric points, the algorithm will in a step-by-step fashion generate a solution that meets the constraints of the original differential equation—this to within some specified error.

But the difference between an analytic and an algorithmic solution to an ordinary differential equation is sharp and it is inescapable. An analytic solution completely penetrates the future or the past; an algorithmic solution acts only over a finite interval of time and space. An analytic solution returns a differential equation to a continuous world; an algorithmic solution,

*The Euler algorithm. John W. Harris and Horst Stocker, *Handbook of Mathematics and Computational Science* (New York: Springer, 1998): 677.

to a world that is discrete. An analytic solution is infinite; an algorithmic solution, finite. An analytic solution commands a polychromatic view of the world; an algorithmic solution, one that is monochromatic. Stability issues remain unremarked and unresolved.

These differences are not only conceptual: they are practical as well. An analytic solution must be discovered; an algorithmic solution, executed. The analytic solution of a differential equation involves successful specification in the dark, but an algorithm throws its human light only over a narrow interval.

The advent of the algorithm has brought a universe of equations within the compass of the ordinary computer; physicists who cannot completely solve Einstein's field equations for general relativity may nonetheless use sophisticated variants of Euler's algorithm to look at toy models of the universe as they evolve in a step-by-step fashion from some specified set of initial conditions. It is, of course, an immensely exciting process to watch, this seeming ability of the algorithm to provide creation with a second go-around. But there is a price to be paid for that sparkling loveliness. A differential equation and its analytic solution belong to one and the same world of discourse; they obey the same rules and moan and mate by the same laws. An algorithm is an alien entity in this world, discrete, finite, moving crabwise through its appointed steps, and forever bearing the mark of its human creator.

CHAPTER 13

An Artifact of Mind

A way of thinking makes a world appear. Whatever we may say, we are still ideologically the party of the physical sciences; like any ideological affiliation, ours involves commitments conceived without justification, the commitments determining the evidence, rather than the reverse, and this by means of a psychological process as difficult to discern as it is to deny. The largest of these commitments, and the one least examined because most tenaciously held, is the belief that the universe is nothing more than a system of material objects. Beyond this system—nothing. A universe of this sort might seem repugnant to most men and women, but many physical scientists have proclaimed themselves satisfied by a world in which there is nothing but atoms and the void, and they look forward to their forthcoming dissolution into material constituents with cheerful nihilism.

An uneasy sense nonetheless prevails—it has *long* prevailed—that the vision of a purely physical or material universe is somehow incomplete; it cannot encompass the familiar but inescapable facts of ordinary life. A man speaks, sending waves into the air. A woman listens, the tiny and exquisite bones in her inner ear vibrating sympathetically to the splashes of his voice. The purely physical exchange having been made, what has been sound becomes what has been *said*; heated by the urgency of communication, the sounds begin to glow with meaning so that an undulating current in the air can convey a lyric poem, issue a declaration of war, or say with terrible finality that it's over. Making sense of sounds is something that every human being does and that nothing else can do. More than three generations of mathematical physicists grew old before their successors understood black-body radiation; the association between sound and meaning is more mysterious than anything found in physics. And we, too, are waiting for our successors.

A MODERN ORTHODOXY

The great body of continuous mathematics that like a living current informs mathematical physics has played little role in the explanation or description of the human (or animal) mind, however much its very existence may express the powers of that mind and so point to a puzzle in prospect. Yet within living memory, the powerful and disturbing image of the human mind as a computational device, something marked, as the digital computer is marked, by the subordination of its routine to a formal system, has initiated a bright new world to rival the old, abstract, and continuous world of the physical sciences.

A. M. Turing's simple model of a computing machine is perhaps the greatest of humanity's *intellectual* artifacts, the prototype having become its own Platonic ideal in an extraordinary metaphysical tour de force. A Turing machine is a device for the manipulation of symbols, and since symbols are abstract, a Turing machine may be realized in any medium in which symbols may be inscribed.

Symbols, note, and thus objects with their own urgent and compelling conceptual identity.

These considerations enforce the conviction that a different organization of the world is in the making, one animated by principles unlike those found in the physical sciences. Whatever a computational system may be, it is in some sense a transcendental object, one that like the human mind itself conveys by some physical means an *im*material something, information, perhaps, the stuff that is stored, recorded, wired, faxed, communicated, exchanged, the impalpable fluid that seeps across international borders and boundaries, the animating discharge that invigorates matter, the essential quality, speaking metaphysically, that gives form and content to the animate world, what is left when the medium of the message is withdrawn from the message itself, the message beyond the medium.

The American mathematician Claude Shannon gave the concept of information its modern form, endowing an old, familiar idea with a perspicacious mathematical structure. His definition is one of the cornerstones in the arch of modern thought. Information, Shannon realized, is a property resident in symbols, his theory thus confined from the start by the artifice of words. What Shannon required was a way of incorporating information into the category of continuous properties that like mass or distance admit of representation by the real numbers. Shannon understood that symbols function in a human universe where things in themselves are glimpsed through a hot haze of confusion and doubt, but whatever else a message may do, its most important function is to relieve uncertainty, "*Rejoice, we conquer*" making clear that one side has lost, the other won, the contest finally a matter of fact. It is by means of this superb insight that Shannon was able to coordinate in a limpid circle what symbols do, the accordion of human emotions wheezing in at doubt and expanding outward at certainty, and the great, the classical, concepts of the theory of probability. A simple binary symbol, its existence suspended between two equally likely states (on or off, 0 or 1, yes or no), holds latent the promise of resolving an initial state of uncertainty by half. The symbol's information, measured in bits, is thus one-half.

Like entropy, information is a temporal quantity, one tied to the passage of time. Before I receive a message, I do not know; afterward, I do. Time must flow for what is hidden to be revealed. The definition of information, and the theorems that it sustains, is addressed to a world in which dice reveal the secrets of their spots at every throw, and coins their face, and letters or e-mail messages their dark, disturbing contents after they have been sent and when they have been read.

Shannon worked for Bell Telephone's magnificent research laboratory, and even though he pursued his thoughts quite without corporate pressure, he was nonetheless interested in the end in bringing order to the primordial model of communication on which the telephone depends: someone is speaking, someone else listening, and there is something else acting to transmit the message between them, the telephone most obviously. The setup that Shannon imagined involved a sender and its source, a signal passing between them in the form of symbols drawn from a finite and discrete alphabet.

Let that alphabet consist of the symbols $S_1, S_2, \ldots, S_n$. The symbols are indexed by the natural numbers: the first symbol, the second symbol, the third, and so on. The symbols are sent from some source by some sender to some receiver and thus to some sendee. Each symbol occurs with a fixed and a priori probability. If there are five symbols in a message, and each symbol is sent by being fished randomly from a hat, then the probability associated with each is one in five. The information resident in a message, Shannon then defined in terms of a simple formula. Symbols are given, these with fixed probabilities. The sum of their logarithms is then taken as the basic measure of information.

$$\textbf{Information} = -\sum_{i<k} \log \textbf{probability } S_i$$

The logarithm is used so that numbers do not explode indecorously and the sign of the sum is reversed to restore the sum to a positive number. But even without the details, the meaning of the formalism should shine through, like the light of a candle seen through gauze. The otherwise recondite property of information has been given a precise mathematical voice in terms of a few very common, very obvious mathematical properties: probability or luck, the logarithmic function, addition, these playing in turn over a synthetic universe of symbols flashing between two sites.

Shannon's definition has led to a rich, productive, and unique mathematical theory, Shannon able to demonstrate that no other definition besides his own quite captures the intuitive properties of information; but the circle of light thrown out by the definition has a very narrow radius. For one thing, the definition is insensitive to the content of a message, treating "*Rejoice, we conquer*," and "*The coin came up heads*" as precisely alike in terms of the information they convey (assuming they measure the same degree of uncertainty).

For another thing, the definition makes clear and compelling sense only as a measure of the information conveyed by a string of symbols. Such strings exist in only one linear dimension; and yet information in the large sense it has acquired as a general term of culture is meant to serve as a surrogate for the complexity of a living organism, a rich throbbing creature in three full dimensions of space and one of time. Just how life crosses the dimensional barrier, passing from one to four dimensions every time an egg grows into a chicken or a chimpanzee, is a very great mystery, one which we are barely able to express, let alone solve.

The invocation of information represents the second leg of a far-reaching three-part conjecture about the nature of mind, one fast becoming a modern orthodoxy. It is computation that explains *what* the mind does; and information that explains what it does it *with*. What remains to be explained is *how* the organism flaunting its mental powers acquired them in the first place. It is here, where the third leg needs to be affixed to the stool of thought, that Darwinian theory has made a surprisingly powerful reappearance, completing what seems to be a remarkable synthetic structure by complementing those otherwise unsteady legs of computation and information.

The human or animal mind has come about by means of the ancient whimsical devices of random variation and natural selection. If there are powerful computational routines at work in vision, speech, locomotion, hearing, touch, taste, memory, pattern recollection, sexual attention, or sexual jealousy, then this is because precisely such routines, having arisen by an accumulation of accidents, proved valuable to the organisms that possessed them.

In the case of human beings, natural selection has played its hand most decisively during the long period of their existence when they were hunters and gatherers, not apes, precisely, but strange simianlike creatures, nonetheless, loping across the savannas, amiably inspecting one another for fleas, living lives that were, so far as I can tell, solitary, poor, nasty, brutish, and short, but, nonetheless, living lives of accumulating sophistication, the invisible but implacable hand of natural selection seeing every stray advantage in the appearance of some new subroutine and seizing on it expeditiously, so that insensibly a creature destined originally to gather up nuts and grunt ineffectively transformed itself into a creature capable of asking the waiter whether the Brie is runny, thus collapsing into one lucid question two divergent points of evolutionary origin.

The plausibility of the picture that emerges is apt to suggest a theory nearing completion, the anonymous cohort forever figuring things out finally having figured things out.

THE INTELLIGENT ARTIFACT

In his later years, Turing managed by some mysterious form of cosmic amplification to have his own internal dialogue become a part of the general conversation; in an article published in October 1950, he asked whether a machine could think, the very idea one that he himself had brought into being, and answered his own question by a conditional affirmation—*Yes, if the machine can fool a human interlocutor into believing it human. No, otherwise.*

Such is Turing's test. Enacting this test has actually become a ritual in recent years, with a row of solemn stalwarts facing a collection of curtained booths, trying to determine from the cryptic printed messages they are receiving—*My name is Bertha and I am hungry for love*—whether Bertha is a cleverly programmed machine or whether, warm and wet, Bertha herself is resident behind the curtain, stamping her large feet and hoping for a message or massage in turn.

Whatever the details of Turing's test, the computational theory of mind is ambidextrous, applying on the left to machines and on the right to human minds, the governing object in either case a program or an algorithm; and it is this utterly unreal sense of having localized an aspect of intelligence in a formal set of symbols that lends the theory its profound, compelling, and deeply disturbing power. Yet the incomplete but tantalizing vision of the mind as a computational object is in its effects retrograde to the great movements of mathematical physics. The conceptual landscape is changed, the world emerging now in a finite whirl of words, integers, or symbols, but *not* real numbers. In the physical sciences, time and space are represented by just those numbers; they have a continuous structure. A computer inhabits a world in which time is represented by the ordinary integers. So, too, an algorithm. Time has lost its pliant seamlessness and moves forward in finite, jerky, integral steps. A Turing machine is inherently discrete. Its fundamental theoretical objects are symbols, not electrons, muons, gluons, quarks, or curved space and time. A stern series of renunciations is in force. No differential equations. No connection backward to the calculus. No world-defining symmetries of space and time. No analytic continuation, as when the laws of nature conduct the physicist from the present into the future. No quantitative miracles. No miracles at all, save for the familiar miracle in which some part of the physical world becomes animate.

Under the new conceptual order, the prevailing direction of thought is altered and reversed, like a current animated suddenly by a change in polarity. Within mathematical physics, things move dissectively *down*ward toward the fundamental objects and their fundamental properties and laws. The universe thus revealed is *meaningless,* its fundamental laws controlling a vast but sterile and inaccessible arena, the whole thing rather like a fluorescent-lit bowling alley, where bowling balls the size of quarks forever ricochet from one another in the monstrously hot and humid night. Down *there,* no human voices may be heard.

Maybe so. But up *here,* let me tell you, things are different. Invoking a rich system of meaning and interpretation, human beings explain themselves to themselves in terms of what they wish and what they believe, the immemorial instincts of desire and conviction sufficient to bring a world into being. It is a world suspended in space by the divine chatter of human voices. A path through the chatter is almost always circular, and not dissective, as when small-town gossip returns red-faced to its source. A man believes that alfalfa sprouts are a cure for shingles; this is reflected in what he says, in what he does—eating alfalfa sprouts—reflected in what he believes and what he wishes, each reflection explaining the one that has gone before, the line of reflections bending back on itself, circular. The power of the circle lies resident in its illumination, the incandescent episodes infused by *meaning*—the force that lights the circle up and the thing that vanishes when in death or in despair the circle lapses. Meaning is alien to physics, arising in the world in response to something impalpable as a thought, a mental shrug.

There is no way to break the circle to reach a bedrock of physical fact. There are no *physical* facts to reach. How could there be? To enter the circle, any purely physical feature of the world must be interpreted and given meaning. Once given meaning, it is no longer purely a physical feature of the world. The conceptual circle is not emptied or evacuated under the computational theory of mind: it is enlarged instead, the formal objects taking their place within the familiar circle, like wedding guests asked to join the wedding dance.

The states of a computer are *representational.* They carry a significance that goes beyond physics; like words, they play a role in the economy of meaning. And meaning appears only in the reflective and interpretive gaze of human beings.

It is at this very point that psychologists and computer scientists are apt to forget the subtle pattern of development that over sixty years ago made possible the advent of the algorithm. A computer is a device; a Turing machine, an intellectual artifact, and both are governed by the logician's double vision by which symbols are simultaneously deprived of and endowed with meaning. From the point of view of the computer, the program that it executes is in the end a series of binary symbols, and the symbols in the end nothing more than a sequence of discrete physical marks. An algorithm manipulates those marks quite without attending to their meaning, thus mimicking the logician in one of his schizophrenic modes.

But no explanation of an algorithm is complete, or even coherent, under the impress of just one mode. To understand what an algorithm is doing, it is necessary to understand why it is doing it. And for this, symbols that have been stripped of meaning must be given meaning anew.

This point is evident in the simplest of devices, a calculator, say. Call twice for the numeral "2" and the machine returns a neoned "4." Considered simply as a physical object, the machine is capable of shuttling between shapes, configurations of light that it realizes by virtue of the way in which it is constructed: it is capable of nothing else; but what makes the charming show of light an *answer* is the fact that someone has been provoked to ask a *question*. Question and answer belong to the circle of human voices. A purely physical process has been invested with significance, those winking ruby lights given form and content as symbols, representations, in fact, of numbers. Whether the representation is made in terms of light or by the modulation of a woman's voice, the process is the same, some feature of the world has been made incandescent.

If this point is evident in the case of a calculator, it is evident again in the case of complex algorithmic structures such as neural networks, or nets. Introduced originally in the 1940s under the generic name of *perceptrons,* neural nets were intended to provide a model of the nervous system itself. Very effective criticisms made first by Marvin Minsky and Seymour Papert seemed to have driven a silver stake through the heart of the program: it turned out that perceptrons could do little of real interest, but rather like an astonishingly robust vampire, neural nets have in the last fifteen years or so staged their own remarkable comeback, often under such names as *connectionism* or *parallel-distributed processing,* and may now be seen regularly leaving their crypt as the shades of evening draw nigh.

The idea of a neural net is simplicity itself. Although the real net is presumed to lurk somewhere in the brain itself, an electronic neural net lives within a digital computer. It consists of a series of nodes or very simple processors. The state of each such processor is determined entirely by signals that it receives from other nodes or processors. Paths between nodes are assigned certain weights, so that a signal passing from one node to another may be multiplied or diminished in strength as it zips or is zapped along.

Each node in a neural net is capable of receiving weighted signals from a number of nodes; typically the node adds its weighted signals, and then assigns them to some function or other, usually sigmoid (or S-shaped, like the sigmoid colon). A very common function is the hyperbolic tangent. Having received its freight of weighted signals, and then summed them, each node sends the packet on to other nodes by means of its function (see figure 13.1).

The result is, of course, a network of nodes, in which the various weights determine, in effect, the response of the system to inputs and so serve as the system's memory.

Beyond the trivial—a single layer of nodes firing blankly at one another—the most interesting neural net configuration is one in which nodes are divided into layers, with each node able to send its signal only to a node above it. Such are the

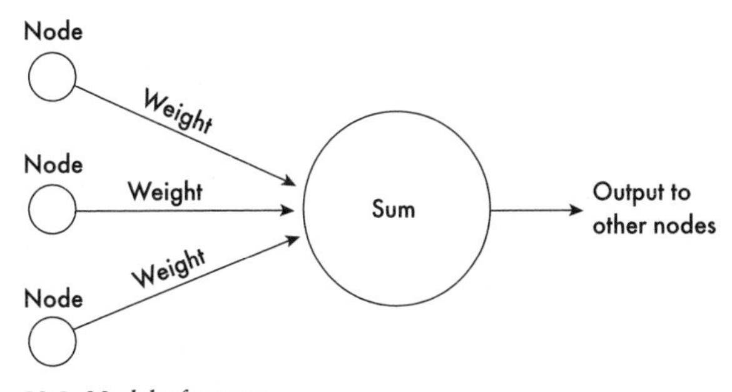

13.1. *Model of neuron.*

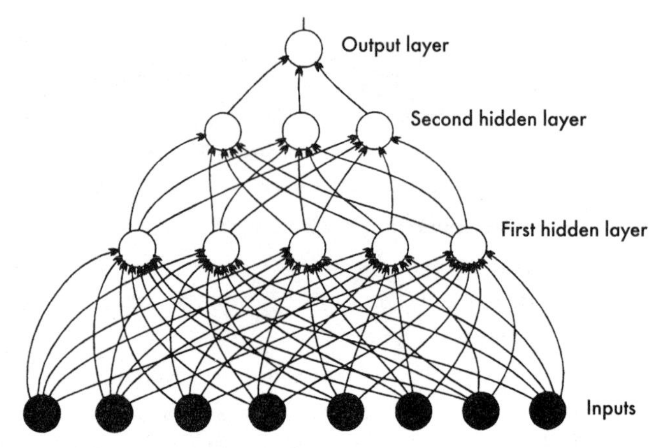

13.2. *Feed-forward neural net.*

feed-forward structures. Layers that receive signals *from* the Great Beyond are input layers; those that transmit signals *to* the Great Beyond, output layers; and in between, there are the hidden layers (see figure 13.2).

In the simplest of such cases, the net is considered without reference to time. Signals are sent as inputs, weighted by connections, summed by nodes, manhandled by functions, and then sent upward. The behavior of such nets can be completely described by two equations. The input signal of a particular node is calculated as the sum of the weights, w_{ij}, assigned the path between node j and node i, and the signals, y_j, being sent to it from other nodes:

$$x_i = \sum_j w_{ij} y_j$$

The input signal is then sent onward by means of its own particular function:

$$y_i = \sigma_i(x_i)$$

The entire behavior of the net may then be simulated simply by solving these equations for every input.

From a mathematical point of view—the only one that counts, really—a neural net functions as a kind of signal processor, where the signals are numbers, which in turn are converted to other signals, and hence to other numbers. Thus, before a neural net may be put to work, it is necessary to convert inputs and outputs into numerical form. With that done, the neural net functions in effect to convert specified sets of numbers to specified sets of numbers. Numbers specified in this way are called *vectors* and so a neural net can be thought of as a device effecting a vector-to-vector transformation.

Although primitive, a simple feed-forward neural net can approximate a continuous function to an arbitrary degree of accuracy by means of an increase in the number of its hidden nodes. This is a mildly marvelous but not an entirely unexpected result; from an algorithmic point of view, neural nets are simply devices for computing computable functions. They are thus bounded at their margin by Church's thesis. Whatever a net can do can be done as well by a Turing machine.

Systems of this sort may be made more interesting and far more flexible by standard techniques borrowed from the theory of feedback control, as when an error is propagated backward through the system in order to adjust its weights. And in a very natural sense, such systems may give every indication of becoming better and better at certain tasks, such as handwriting or pattern recognition, speech synthesis, grammatical analysis, and the like. In such systems, a learning algorithm is incorporated in the system's design. Nothing mysterious here; in fact, nothing more than an old engineering staple (on display in an ordinary thermostat) in which the system's output is corrected with respect to some fixed mark and its weights automatically adjusted until the system gradually converges toward some specific goal: getting the handwriting right, distinguishing nouns from adverbs, recognizing the face said to be typical of multiple miscreants.

The great advantage of neural nets lies in the superb ease and flexibility by which they can be *trained* (an interesting choice of words); given a random assortment of pictures, some depicting multiple miscreants and the rest ordinary women, a neural net can be taught to converge on the miscreant's features by means of examples alone, the process undertaken quite without any specification of detail on the part of the grim police inspector or his outraged victim. An algorithm executing precisely this task is easy to imagine. First the photographs must be recast in digital form, and then presented to the net as numbers; next—the algorithm undertakes the following steps:

PRESENT the net with random photographs
EVALUATE the output
IF the net misidentifies a photograph as a multiple miscreant,
  adjust its weights
IF not, do nothing
REPEAT until the net successfully identifies multiple
  miscreants as multiple miscreants
STOP

There is no question that there are interesting and important intellectual acts that can be embodied in neural nets, and no question either that their potential was dismissed prematurely. Nonetheless, neural nets, are no less deeply involved in the intentional circle than any other computational device. This is a point widely misunderstood. Thus the Harvard biologist Francisco Varela draws an unfortunate distinction between the cognitive theory of mind and the theory he imagines embedded in neural nets, a theory that he has come resolutely to champion. The first requires he believes, but the second does not, an entirely supererogatory apparatus of symbols and representations. Neural nets require neither.

But this is to confuse the architecture of a system with its interpretation. A neural net is a computational device, one passing signals to signals. So, too, an ordinary digital computer. The architecture is somewhat different, the digital computer proceeding serially, and a neural net proceeding by means of parallel processing, but the symbolic interpretation of both neural nets and computers has nothing to with their hardware; *that* lies at another level entirely.

A neural net that has learned to recognize a pattern has done something that goes beyond what it has done. A pattern is only a pattern by reference to some scheme of human interpretation. The input to a neural net may be a signal; so, too, its output, but signals become symbols only when the net comes to something of human interest. Imagine a neural net given photographs or digital images of thousands of human faces and trained over time to pick out the distinctive physical characteristics associated with certain criminal types. (It is by no means clear that there are *any* physical characteristics associated with crime, but this is a thought experiment.) Such a net might explain the fact that *we* can spot the tormented face of a multiple miscreant given only a few scant clues—the ridge of bone above her eyes, tensed and narrow volutes, drooping earlobes, a certain shiftiness in her gaze. These are features that we may observe but cannot necessarily specify. But what we have done, we who can pick out the deviant from among the decent, decorous, normal women, goes considerably beyond looking at a series of pictures and uttering the tense locative: *Lo, a multiple miscreant.* In order to identify a woman as a miscreant, we must mean what we say. Mean it in English (or any other language), and mean it with the full force of discovery and dismay. These are acts well within the intentional circle.

What holds for us holds as well for neural nets, the device converging on the face characteristic of a serial rapist, habitual shoplifter, or compulsive gambler having done something of interest only when its output—a series of numbers, after all—is given content as a representation of a human face. Without such a representation, the entire exercise collapses into an exchange of signals, both signals a part of the physical universe and both quite without meaning.

Much the same thing is true, of course, for an actual digital computer, which is purely a physical instrument operating on bits of wire, chips of silicon, and surges of current. A physical computer contains neither symbols nor representations. These *we* give to *it*, endowing its states with meaning or significance. Without the endowment, there is no explaining its behavior satisfactorily—why, for example, when prompted to find the square root of 36 it responds with a shape representing the number 6, and not some other shape. The laws of physics provide an explanation for its behavior that is necessary but hardly sufficient to account for what it does, just as the laws of chemistry provide an explanation for the creation of plastic that is necessary but not sufficient in order to explain its existence. If we wish to know *why* the computer exemplifies a particular physical pattern, current surging through its logic gates in just a particular way, then we must look to the program it instantiates, the design that it implements, and the intelligence that it represents.

Attempts to resist these demands are unavailing. There is no way to see in a purely physical system the axioms of a mathematical structure without first *seeing* them there. Even if by means of some elaborate demonstration one could show that there is an uncanny and elaborate correspondence between the physical shapes instantiated by a computer and the number system of ordinary arithmetic those shapes are meant to represent, the demonstration would be beside the point. A *corre-*

spondence is again something that human beings notice and then establish.

ON ALIEN SHORES

The idea that the human mind is essentially a computational device commends itself by virtue of an unanswerable question: What else could it be? But whatever the rhetorical merits of the claim, the idea that the mind is *nothing more* than a computational device subordinates itself to an unavoidable challenge: If that is so, what precisely are the computations (and so the algorithms) involved?

Human beings have beliefs, they make claims to knowledge, they are driven by desires or find themselves vexed by trifles, they are by turns angry, petulant, self-absorbed, self-indulgent, or forgetful; it is not only belief and desire that require a computational account, but attending, noticing, discovering, hoping, finding, looking, peering, peeping, mastering, learning, acquiring, demanding: a man may be angry and rational or irrationally angry; he may wish against wish or hope against hope; his mind may be turbid, lucid, clouded, or calm; he may be anxious, satisfied, relaxed, or vengeful, often all at once; his memory of the events of 1927 may be excellent or scattered; his thoughts confused or coherent, his nature at war with itself or harmonious; he may laugh at the devil or take his calls cheerfully on a cellular telephone; he may fear dishonor more than death, or the reverse; he may entertain bids and make decisions, one after the other, like a string of firecrackers going off; he may choose to meditate on the form of the Eternal One embodied in a shower curtain, or he may repose himself in memory, living amidst the past; and over all these things that a man may do, or the states in which he might enter, the computational theory of mind must play, showing in detail just how it is that what human beings do in the world of human thought and action may be captured by a computable function or certain lines of code, these signaling a computation in prospect or in the process of execution.

■ ■ ■ ■ I was reminded of this entirely unoriginal point by an oddly complete memory. It was more than twenty years ago; I had been teaching something or other at a small school in the Pacific Northwest that had somehow managed to permute the letters of its own name so that in recollection, I think always of the place as the University of Pungent Pork. A quick description, no more than a snapshot: lush lawns, red brick buildings, low clouds, the cold gray choppy waters of Puget Sound off in the middle distance, mountains somewhere else, one monster—Mount Rainier—suddenly looming in the sky on clear days with a kind of repulsive majesty.

It was at UPP that I made the acquaintance of the mathematician Leo Rubble, a very brief obituary notice in the *Journal* recounting the fact that his last hiking trip was followed by his first heart attack recently reminding me that for a time he and I had shared an office in Rummelhart Hall. He was short and genetically destined to be soft and round, but by virtue of a rigid and utterly inflexible program of weight lifting, he had managed to reshape his pudgy torso so that he gave the appearance of being fashioned from some material denser even than human muscle; molybdenum, perhaps. I had seen him bench-pressing more than four hundred pounds.

One day, he told me the following story. As far as I can tell, she had entered his life like a sudden shaft of sunlight seen through clouds; that very afternoon, driving in her little blue Toyota, they crossed the Tacoma Narrows Bridge and checked into a motel called appropriately enough The Narrows. The room was taken up almost entirely by an enormous king-sized bed, with a brocaded coverlet. There was a picture of a sad-eyed señorita on the wall, Leo recalled, done in pink and pastels.

He had been aware of her from the first day of class, her lavish blonde beauty simply spilling from her like water from a careless fountain, the light catching at the fine fuzz along the splendid lines of her temple; but he had known her only since that morning, when, with perfect confidence, she cornered him in his office.

"You bother me," she had said to him. And to me Leo added, "Can you imagine?"

Could *I* imagine? Everyone on campus had noticed her brilliant blonde beauty; she was older by perhaps five years than the other undergraduates and married as well to an Army captain stationed at a nearby base. Her name was Ann Preval and *she* bothered everyone, perhaps because the rose of her beauty seemed to fold over a pistil of sadness and despair.

Now through some concatenation of circumstances that I could not quite fathom, Leo Rubble found himself in the small alcove of the motel room separating the bedroom from the bathroom.

He encircled her waist, he told me, and kissed her lips, his heart thumping.

"I love you," he said.

She said, "And I already love you."

The afternoon passed; on the way back to campus, they drove through the low layer of clouds that had covered Puget Sound.

She failed to appear in class on Wednesday and again on Friday.

"I was going out of my mind," Leo said to me with the kind of terrible earnestness that is so often an excuse for an unacceptable innocence. He had a wife, of course, and a family, and the usual detritus of an academic career, and he was, as well, a modestly talented mathematician, a specialist in differential topology at a time when differential topology was very popular, the sinuous curves and rills of his subject having just possibly laid the foundation for what was plainly an enormous susceptibility. He had a lot to lose, and although I am not certain of this, I believe that in the end he lost it all.

That evening, after dinner, he drummed his fingers on the table restlessly. The Rubbles lived north of UPP on a wooded island.

"You're restless, go for a walk, Sweetie," his wife had said reasonably.

"Maybe I'll just drive around."

A fine light rain had begun to fall by the time Leo Rubble had crossed the bridge to the freeway, the enormity of the deception he was about to undertake brought home to him by Linda Ronstadt's voice. "She was singing 'You're no good,'" he said ruefully, "and I just knew she was singing about me." Years later when I met Linda Ronstadt herself, I confirmed that the man she was singing about was *nothing* like Leo Rubble.

My poor Leo had only the vaguest idea where Ann Preval lived, the information that he had committed to memory consisting of a series of arboreal turns—Oak, Chestnut, Elm, Cedar—ending finally in Monterey, which was a cul-de-sac; he knew that her husband, the captain, was on a training flight.

He turned off the freeway at the first Westwood exit, which gave onto a desolate boulevard, a Pizza Hut spreading a sad penumbra of light into the misty night air; farther on there was a closed used car lot; and then still farther on an open garage. Looking at the map, it had seemed simple, right on Oak, then right again on Elm, and left at Chestnut, thereafter, a left on Monterey; but when he had finished making the turns, he found himself back on the broad boulevard leading to and from the freeway.

He pulled into the garage for no better reason, I suppose, than it seemed something that he might do, and draped the map over the steering wheel.

A heavyset elderly man—the house mechanic or garage owner evidently—emerged from the interior of the garage and walked sedately to his car.

"Help you?" he asked. He had a gentle Parkinsonian tic that passed over his face as he spoke.

"I'm trying to find someone," Leo Rubble said.

The mechanic spread his hands apart to indicate the hopelessness of it all.

"She lives on Monterey Lane."

"Boulevard or lane?"

Leo shrugged helplessly; the distinction had never arisen. "She's blonde, kind of statuesque." He carved the air in front of the steering wheel with his hands in order to indicate her statuesqueness.

"Blonde, you say? Tall girl?"

Leo Rubble nodded.

"The one you're talking about drive a little blue Toyota?"

"That's the one."

"She lives over on the boulevard, not the lane," the mechanic said, and the curious thing was that as he said it the faintest impression of a sly smile seemed to play across his heavy leatherish face, the deep folds along the cheeks waggling upward.

Holding the top of the map with one hand, he pointed to Monterey Boulevard, which did, in fact, lie just beyond the forest of Oak, Elm, and Chestnut, but at right angles to Monterey Lane.

"Can't miss it," he said, pointing out the steps with the stub of his index finger. "But no use you trying," he added, the same infinitely subtle, infinitely sly smile never leaving his face. "I seen her go out before."

Leo Rubble folded the map slowly and placed it on the seat next to him.

"Do you for anything else?" asked the garage attendant.

Leo Rubble shook his head; after allowing his face to be

modulated by another gentle tic, the mechanic limped slowly back to his open office.

"What happened then?" I asked.

"Nothing, nothing happened. That's the story. She never wanted to go with me again. I think she's in Europe somewhere."

As I said: an odd little story. Let's see the programmer who can write the code capable of seeing what Leo Rubble saw.

■ ■ ■ ■

THE INELIMINABLE

Streaming in from space, light reaches the human eye and deposits its information on the stippled surface of the retina. Directly thereafter I see the great lawn of Golden Gate Park; a young woman, nose ring twitching; a rosebush; a panting puppy; and beyond, a file of automobiles moving sedately toward the western sun. A three-dimensional world has been conveyed to a two-dimensional surface and then reconveyed to a three-dimensional image; and this familiar miracle suggests, if anything does, the relevance of algorithms to the actual accomplishments of the mind, the transformation of dimensions precisely the kind of activity that might be brought under the control of a formal program, a system of rules cued to the circumstances of vision as it takes place in a creature with two matched but somewhat asymmetrical eyes.

It is precisely work of this sort that was undertaken by the late David Marr at MIT.* The process of vision begins, on Marr's theory, with the retina. Light strikes this two-dimensional surface, the result a pattern reflecting the intensity of light at different points. Such is the input to the brain's

*David Marr, *Vision* (San Francisco: W. H. Freeman, 1982).

visual system. Computations then transform this intensity array until the brain recovers a three-dimensional representation from its original input. Its elements are not letters or numerals, but what Marr calls visual primitives. Just as their name suggests, these visual primitives embody visual information in attenuated form, with lines standing for edges and objects such as cones or other solids for their volume. These three dimensional representations are the output of the brain's visual system.

Given the enormous sophistication of Marr's work, one thing must strike readers of his book as curious. There is no entry for *seeing* in its index. And this is evidence of a striking conceptual difficulty, one that has its origins in the ambiguous nature of representations as physical objects *and* as visual objects. If representations are considered simply as formal objects, they cannot be seen, if only because formal objects have no obvious visual significance; indeed, they have no significance whatsoever.

Should representations then be understood as visual primitives which, in fact, designate or otherwise describe the visual field? Indeed, the language of representations and images is general throughout the cognitive sciences, the mind storing the stuff in various places and hauling down a representation or two when the need arises.

But wait a minute. *Representations*? *Images*? As in something *seen*? *In* the mind? But my dear Doctor, *images* seen by *whom*? And just *who* is doing the representing?

The computational theory of vision was to provide an account of vision; it was, moreover, to provide that account in computational and so in physical terms. But evidently when the computation has run its course, the visual system having deposited a representation within the brain, the brain is called upon to do something suspiciously like seeing. How else does it make sense of a representation?

It is, of course, entirely possible that what is suspiciously like seeing is not quite seeing at all, but something weaker in nature; and Marr does suggest that in interpreting a representation, the brain *recognizes* certain visual features, *compares* them to other visual features, and in general carries on a number of *cognitive* activities below the threshold of vision itself. On this view, seeing something is a complex activity which may be decomposed into its parts. Needless to say, this elaboration of detail hardly constitutes an improvement in principle. Whatever the aspects of vision, if they are understood strictly in formal terms, they are apt to be uninformative, the magic moment in which a visual primitive is *recognized* no more accessible to computation than the magical moment in which a visual primitive is seen. If not accessible to computations, these primitive aspects of sight must be understood as activities every bit as mysterious as sight itself. An analysis of this sort may well be of great scientific value, if only because it might well succeed in explicating one mysterious cognitive act in terms of a great many other mysterious cognitive acts. But one thing it does not offer is an escape from a circle of mental concepts. These, it would seem, remain ineradicable.

These problems reverberate with a loud, flat embarrassing *bang!*, their ideological innocence utterly at odds with the very real sophistication of the various theories they subvert; despite every assurance to the contrary, the fact remains that a curious and inexplicable group of troll-like figures pops up regularly in the cognitive or computational sciences, doing the seeing when sight is under investigation, or smelling the smells, or handling the other cognitive chores with an easy and horrible familiarity.

It is an old affliction to which I am calling attention, one that infects Freudian psychology as well, the Freudian ego a character about as assertive, fractious, and demanding as the character it is meant to explain, so that on occasion, elderly analysts report, the patient and his ego appear to be rancorously contending for possession of one and the same scruffy piece of mental ground.

A stubborn and persistent infirmity is a sign of conceptual confusion and not merely a mark of carelessness. Is the mind computational? It is. Does it proceed by an application of determinate rules? It does. Very well. Consider this: At the conclusion of its computations, the mind bursts into consciousness, a vivid and light-enraptured awareness of the world. *I* open my eyes and my eyes are filled, the simple achingly miraculous act conveying each and every time it is performed the doubled nature of experience. My eyes are *filled*. There is a panorama to which they are partial; but it is *my* eyes that are filled, my experiences possessing both an experiencing subject—*me,* as it happens—and the *contents* of that experience, the scene and the mysterious stranger surveying it bound inseparably together as fragments in a figure. But if consciousness is ineffably divided and yet ineffably complete, computations are by their very nature sequential, one thing proceeding from another, as in a chain reaction. The persistence in theory of a certain embarrassing imbroglio, the mind suddenly opening an arena in which images are thoughtfully examined, or representations mysteriously made to represent, is evidence of the enormous difficulty in accommodating the essential nature of consciousness within any sequential or procedural view of the mind's operations. If one of my troll-like assistants is doing the looking down there, it is hard to see how the experience of vision has been explained; and if no one is looking at anything, hard again to see how consciousness has been accommodated.

Whatever the difficulties, most philosophers have remained rhetorical materialists. Like members of a lodge, they are unwilling to confide their doubts to one another. At lodge meetings, they sing songs to keep their spirits up. Still, it is consciousness that is now on everyone's lips, and not materialism. A great many people seem to have concluded that they have but to open their eyes to fashion a rebuke to prevailing views. And they are right. Philosophers are confounded—by their irrelevance, if nothing else. A few have been seen administering a number of discreet kicks to what appears to be the corpse of dualism: *Get up, you fat fool, I need you.* Employing an argument prematurely discarded by logicians, the distinguished mathematician Roger Penrose has concluded that consciousness could not be computational.* A reformation of quantum theory is required to set the matter right, the transmutation of thought into action taking place in the microtubules of the cell. No point on the compass of possible positions has actually gone unvoiced. Unorthodox quantum physicists have argued for the ubiquity of mind throughout the cosmos, the stuff turning up everywhere, with even the very atoms having a say-so in the scheme of things.† An enterprising philosopher—who else?—has concluded that the problem of consciousness must forever be insoluble and, flaunting his ignorance majestically, has made that discovery the foundation of a far-reaching philosophical system.

Do *I* have anything better? No, of course not. Let the last word, then, come from the Greeks. "*You could not discover the limits of soul,*" Heracleitus wrote, "*not even if you traveled down every road. Such is the depth of its form.*"

*Roger Penrose, *The Emperor's New Mind* (New York: Oxford University Press, 1989) and *Shadows of the Mind* (New York: Oxford University Press, 1994).
†Nick Herbert, *Elemental Mind* (New York: Plume, 1993).

CHAPTER 14

A World of Many Gods

The calculus and the body of mathematical analysis to which it gives rise is the great idealization of Western science, the moment of its creation marking a profound division of human experience. From within its crabbed formulas comes the master plan of equation and solution that makes possible the physical sciences. Yet the investiture of mathematics in things or processes weakens as one moves up the intellectual chain of command, a curious and disabling but nonetheless incontrovertible fact. Material objects on the quantum level may be explained as roiling waves of probability; the governing equation has its roots in the calculus. The attempt to discern in the structure of *biological* objects—protozoa, rock stars, human beings—the outlines of a coherent system of continuous mathematical thought has been a failure, the miracle of mathematical physics not repeated, the bargain not struck again.

The conceptual landscape of biology is desultory and distinct, a near range of ancient folded foothills against the mathematician's high alpine peaks, the biologist employing a scheme immeasurably simpler than the one adopted by the mathematical physicist. It is, that scheme, discrete, finite, and combinatorial. No mathematics beyond finger counting. Molecular biology would be comprehensible to someone who knew nothing of modern science, continuity or the calculus, and could reckon only to powers of ten—a Harvard graduate, say. Living systems may be understood in terms of the constituents that make them up. Of these, there are only finitely many. The dissection complete, what remains is a master molecule, DNA, that functions as a code (although one without an obvious plain text behind it) and the complicated proteins that it organizes and controls. No continuous magnitudes; no real numbers; no rich body of mathematical analysis. No laws, not in the sense in which physics contains laws of nature; no fantastically pregnant, compressed, and quantitative apothegms; no place where the knot of nature is so tightly wound that it may be touched directly by a mathematical formula.

Despite the frequently vulgar language in which they are expressed, the concepts that animate molecular biology are old, familiar, haunting, and obscure—*complexity*, *system*, *information*, *code*, *language*, *organization*. Many of them affirm a message already known: One generation passeth away and another generation cometh. Some of them may be combined at will, as in *organized complexity* or *complex organization*, an interesting example in which combinatorial processes are expressed by combinatorial words. They are very often magical, DNA, in particular, functioning as a kind of biochemical demiurge, something bringing an entire organism into existence by a process akin to a casting of spells. Often they are inconsistent, the role attributed to DNA at odds with the obvious fact that the information resident in the genome is inadequate to specify the *whole* of a complex organism. They play, these strange and seductive concepts, no role in physical chemistry or even biochemistry. Like a rubber band under tension, they seem always to snap back to some earlier way of describing life, one in which purpose and design come prominently into focus. How else to describe the eye without mentioning that it is designed in order to see? They often seem to mark the very margins of our own intellectual inadequacy. Nowhere in nature do we *ever* observe purely mechanical forces between large molecules giving rise to self-contained, stable, and autonomous structures such as a frog or a fern, something able to carry on as a continuous arc from first to last, a physical object changing over time, but the *same* object in every case, some set of forces endowing its identity with permanence so that variations remain bounded and return inevitably the figure to the place from which it started. Nothing beyond a living system exhibits this extraordinary combination of plasticity and stability, a fact we are barely able to describe and entirely unable to explain.

Molecular biology is insusceptible to the great idealization that marks the physical sciences; and what is more, it seems retrograde to the deepest metaphysical assumptions that the physical sciences make, assumptions that have themselves passed directly into the life of popular culture. The world, the physical sciences affirm almost with one harassing voice, is physical and not spiritual, numinous, or mental. It is a world of matter. The doctrine of consideration and the bright bubble of consciousness are illusions alike. Reality contains only atoms and the void. These brittle declarations sound like rifle shots; they prompt the edges of the human heart to curl in anticipated dismay. But if by physical, physical scientists mean concepts *like* the concepts found in physics, then the conclusion is irresistible that molecular biology is not a physical science at all but a discipline struggling to express the properties of living systems in a vocabulary and by means of a set of concepts unlike anything needed anywhere else.

Intellectual monotheism expresses the conviction that in the end one magnificently unified system of theory and description will account for the whole of the observable universe, the great vault of space and time *and* the miracle of the mammalian liver, a scheme of thought generous and powerful enough to reveal a *single* hand behind the world's varied handiwork. During the 1930s and for a time thereafter, intellectual monotheism came officially to be expressed as a commitment to the unity of science. An ecumenical organization was established to promote its adoption as a creed. The faith has been propagated by impulses similar to those that prompted the people of the ancient world to see in their various gods and goddesses *aspects* of a single, commanding, and inscrutable deity. And yet if the instructions delivered are to *look* at what is in fact the case, the conclusion seems irresistible that one god, like dark Pluto, rules the quantum underworld; quite another, like Pan, perhaps, the biological macromolecules.

Physicists reject this tolerantly polytheistic view, of course, but physicists can say only that the laws of physics are controlling and in the end everything will be made clear. This is what they *always* say. It is their destiny to say it. In truth, the grand vision of all of human knowledge devolving downward toward mathematical physics is no longer taken seriously, even by physicists who take it seriously. "The most extreme hope for science," Steven Weinberg has written, "is that we will be able to trace the explanation of all natural phenomena to final laws and historical accidents." Machiavelli dubbed the inexplicable adjurations of fate *fortuna,* a word that communicates a certain grave mockery. How has mathematical physics informed the anxious human heart if the explanation for the way things are involves an appeal to the fundamental laws of physics *and* something akin to a Neapolitan shrug? Chiliastic physics, it would seem, has entered the same depressing defile inhabited for so long by the Darwinian theory of evolution, the two theories palpating one another's padded shoulders and exchanging damp handshakes down there in the dark.

It is here, at the meeting place of melancholy, mummeries, and mystery, that the algorithm has again made its advent.

THAT KIND OF STORY

▪ ▪ ▪ ▪ There are dreams in which everything is seen, but nothing is understood. A train pulls into an empty station, chuffing in the crisp morning air. A clock solemnly strikes the hour. The double doors open to allow a hunchback with glittering black eyes to exit. No one else leaves the car, but on the platform, there is suddenly the smell of roasting chestnuts.

This is *that* kind of story. A molecular biologist and mathematician at the University of Southern California, a gentle dreamy oddball by all accounts, conceived the idea—but *wait,* let me explain the problem first.

The cast of characters involves a traveling salesman and seven cities. That salesman first: sallow skin, pouches underneath his eyes, his suit from the rack at JC Penny's, Arthur Allan Waterman, to give him a lovely liquid name, one somewhat at odds with the Broderick Crawford head attached directly to his Lee J. Cobb torso. I think of Waterman gruffly endeavoring to disburse rubber goods: tub hoses, earplugs, shower caps, noseclips, bath mats. (Traveling salesmen seem to have disappeared, but Waterman's real-life role model occupies a cupboard in memory's mansion; he was a Fuller Brush salesman, a German refugee trudging the streets of northern Manhattan, explaining in fractured English—all sibilant hisses and misinformed diphthongs—the advantages of certain bristles and brushes to thin-lipped Irish housewives who, given the plausible evidence standing in front of them, had come to the lunatic conclusion that not only were all Jews German but that all Germans were Jews.)

Starting at Witten, Iowa, Waterman's problem, poor schnook, is to visit Wittless (so called because of the mysterious disappearance in the early 1900s of a family named Witt, whose tracks leaving town were observed to have vanished in the prairie snow), Grainball City, Sad Sac, Amblot (the site of three corn-husking plants), and Waterloo just once, while concluding his trip at Wapping Falls.

For the moment, Waterman is sitting at *his* kitchen table in Des Moines, rubber goods piled on the counter, endeavoring by means of the plain pencil with which he is rubbing his ear, to organize his schedule and thus his life. An unfiltered cigarette lies in the ashtray, smoke curling. There is a double aspect to his problem. The first is purely theoretical. Waterman needs to determine how many *possible* routes might convey him from city to city. He leaves off rubbing his ear and rubs instead the throbbing artery in his temple. It is the sort of question forever asked in high school, and he was the sort of student (rather like me, actually) who sat stiffened incompetently in his chair as all around students stretched their straining arms into the air, saying, "*Ooooh, call me, call me, I know.*"

Let me rehearse the line of reasoning that Waterman and I both missed. If there are two cities, there are 2×1 routes between them: such is the old back and forth; if three, then $3 \times 2 \times 1 = 6$ routes, each route now connecting *three* cities, *not* two. Four cities require $4 \times 3 \times 2 \times 1$ or 24 routes, each route connecting *four* cities. And five cities...

It is possible to continue this exercise indefinitely, but the mathematician in all of us—the mathematician in Waterman and certainly the mathematician in *you*—wishes for a general rule connecting the number of cities, whatever it happens to be, with the number of routes. I shall provide the rule myself, like the obliging maître d' that I am. If there are n cities—where n may designate any whole number whatsoever, 1, 2, 3, or 4,327—there are *n*! routes between them, the same *n*! used as an illustration in chapter 12 amazingly having agreed to do useful work in the workaday world (rather like a lipstick model seen scrubbing dishes). With seven cities on the wheat-filled plains, there are 5,040 possible seven-city routes among them, a flickering arrow flashing from one city to another in the prairie night.

But against the number of possible routes, Waterman rises to remind me, there are those that are actually at his command. The possible routes depend only upon a mental map in which cities are linked by the mind's eye. The real routes are otherwise. They are the routes traversed by yellow rural buses, dirt-encrusted mud flaps protecting their balding tires, or by commuter airlines with names like Prairie Sunrise, the grimly smiling hostess sneaking an anxious look at the thumping propellers as she escorts her tired charges up the slush-filled ladder to the plane's ominously vibrating interior. It is the real routes—*these* routes—that are Waterman's concern.

And ours.

Now I must flash-forward Waterman from his kitchen table to the forlorn bus station in Witten, all faded WPA grandeur and reeking urinals. His hanging bag held over his shoulder by two tense fingers, Waterman contemplates the intercity schedule on the station wall, this listing connections *between* cities, the glyph at the intersection of Wittless (left) and Grainball City (top) signifying passage *from* Wittless, where the town's only diner is presided over by a waitress named Doris, *to* surprisingly prosperous Grainball City, its citizens proud to have sold soybeans to Somalia. An X signifies that there is no connection between cities (see figure 14.1).

The schedule signifies, as all such documents really do, the harsh nature of a world in which men wake from uneasy sleep, dress in the darkness, their families snoring in the winter night, and trudge out alone onto the ice-slicked streets.

Departs	Arrives						
	Witten	Wittless	Grainball City	Sad Sac	Amblot	Waterloo	Wapping Falls
Witten	X	🚌	✈	X	🚌	X	🚌
Wittless	🚌	X	🚌	🚌	X	X	✈
Grainball City	✈	🚌	X	X	X	🚌	🚌
Sad Sac	🚌	X	X	X	✈	✈	X
Amblot	🚌	🚌	X	X	X	X	X
Waterloo	X	X	X	✈	🚌	X	X
Wapping Falls	X	X	✈	🚌	X	X	X

14.1. *Intercity transportation schedule for cities that Waterman must visit.*

The question whether there is a route that Waterman can take that *begins* at Witten, *ends* at Wapping Falls, and allows him to visit each of the remaining cities *just once* thus has an undeniable urgency denied to more abstract problems in mathematics or philosophy.

Well, is there?

■ ■ ■ ■

CALL FOR COMPUTATION

The question seems made for the computer and so seems made to be resolved by an algorithm. And it is easy enough to write an algorithm that will solve it:

GENERATE *random paths through the schedule*
KEEP *any path that starts at Witten and ends at Wapping Falls*
DISCARD *all other paths*
KEEP *any path that enters exactly seven different cities*
IF *no path enters exactly seven different cities*
GENERATE *random paths through the schedule again...*

These six lines, which are easily expressed in BASIC, *do* in time serve to settle Waterman's scheduling problem, but there is something distinctly unnerving about the instructions they embody. Generate *random* paths through the schedule?

You mean by, like, fishing the names of cities from a hat and all?

That *is* what I mean and it *does* seem an oddly undisciplined task for a computer to undertake, suggesting an unsuspected streak of frivolity in a friend hitherto known for unassailable sobriety. The program may result in a satisfactory solution to Waterman's problem at first fish; but Waterman is himself as capable as anyone of fishing names from a hat; and what is worse—*far* worse—is the somber fact that there is no telling for sure that those fishy paths may not forever convey poor Waterman centrifugally toward Grainball City and nowhere else, or force Waterman aimlessly to loop between Sad Sac and Wittless, or even send Waterman inexorably toward Amblot, where from his hotel window he may observe with bleak frustration the somber outlines of the local corn-husking factories.

These circumstances have a curiously modern air, this despite Waterman's old-fashioned profession. The relationship between routes and cities is dangerously unstable. As cities are added to Waterman's schedule, the number of routes among them blows up. (In fact, by Sterling's formula, $n!$ is asymptotically equal to $(n/e)^n\sqrt{2\pi n}$ and so, exponential in n.) Such is *combinatorial inflation,* a kind of comic cousin to the cosmic inflation astrophysicists believe blew up the universe directly after the big bang.

As one might expect, this problem and its cognates are of exceptional interest to computer scientists, and their analysis opens to what for the moment is the deepest problem in algorithmic theory. Theoretical computer scientists, like everyone else, organize their concerns by making lists. In this case, their list comprises only two items and affords the theoretician a panoramic view of possible problems. A problem that can be *solved* in polynomial time belongs to what is called the complexity class **P** (polynomial); and it belongs there if its solution can be reached in a bounded number of steps.

The idea at hand is simple.

A polynomial function has the form

$$f(x) = a_n{}^x + a_{n-1}{}^{x-1} + \ldots + a_1x + a_0.$$

A computer scientist requires ten steps or so to state Waterman's problem. The problem lies safely in **P** if there is some bound on the function $f(x)$—a number beyond which $f(x)$ does not go—such that its solution can be reached in no more than $f(x)$ steps. If, for example, it takes 10 steps to state a problem and 10^3 steps to find its solution, the problem is polynomially bounded (by the polynomial $a^3 = 1{,}000$).

Within the complexity class **NP** (nonpolynomial), on the other hand, reside problems whose solutions can be *verified* but not necessarily found in polynomial time. Waterman's problem plainly lies here. It is easy to check whether he has a solution if he has a scheduling scheme. Does the proposed scheme actually convey Waterman from Witten to Wapping Falls while touching on the remaining cities just once? Just look! If it does, the scheme is a solution. If not, not. The hard part is finding the right scheme in the first place; *that* may take forever.

It is the relationship between **P** and **NP** that is of great conceptual and practical interest. Plainly, **P** denotes the world's tractable problems, and **NP**, those problems that are not tractable; and what theoreticians and everyone else would wish to know is whether there exists some method for reducing intractable problems to those that are tractable. As theoreticians say, at close of day, does **P** = **NP**? Can problems whose solution may easily be verified be reduced to problems whose solutions may easily be found? For the moment, no one knows.

Such are the concerns of theoreticians. But think about this. Waterman needs to manage only seven cities. Modern airline schedules typically coordinate *hundreds*. Even the largest and most powerful computers are incapable of dealing with numbers of this size.

There is, so far as we know, no way of getting around combinatorial inflation or reducing every **NP** problem to one that is **P**. It is the low step in the garden walkway that trips up man and machine alike. The low step is not inevitable, like death or taxes. Some bright toddler, gurgling in a crib in Seoul or

Calcutta and dreaming of becoming a big shot in the year 2020, may have the solution to the scheduling problem trembling on his or her budded nerve endings. For now, that step, like Oprah Winfrey, just happens to be there, irritating and inescapable.

But there is a next-best thing. Speed and size are often substitutes for smarts, and if no one can quite figure out how to solve scheduling problems directly so that the computer ingests the data and promptly spews out the solution in polynomially bounded time, it might be possible, the sunny voice of common sense affirms, to go through the requisite random searches rapidly, and by means of computational techniques that take on a great many random searches at once.

It is here that the familiar immerses itself in the improbable, as in a dream; for what Leonard Adelman did was to encode Waterman's scheduling problem into the wet works of molecular biology, allowing the ancient and intelligent machinery of life to get busy in his practical affairs.

LITERARY LIFE

It often happens that a scientific achievement comes about simply because someone has noticed the obvious. (This is an observation that is obviously its own best friend.)

One hundred years ago, it was possible to imagine that a living organism resolved itself into fairly simple structures, organisms composed of organs, organs composed of tissues, tissues of cells; and cells, nineteenth-century biologists often pictured as nacreous globules, little pearls with no particular or overwhelming degree of internal complexity. But why place things so far in the past, when as recently as 1955, Mr. Redenheffer, who, in hours away from the gymnasium where he coached high-school girls in volleyball, taught a course called Blood and Urine Analysis in which the standard bugs—*Es-*

cherichia coli, Streptococcus pyogenes, and the like—would be discussed as if they were jellylike droplets of only a few moving parts, easily classified by the stains they induced or the diseases they caused. Today we know that at the molecular level, living systems are fantastically complex, Redenheffer's blobs, luminous jeweled escutcheons.

Viewed under low-powered magnification, the bacterial cell *does* look like a drop of jelly, glistening, uninteresting, and slimy—by pressing the heel of my palm gently into my eyes, I can recover Mr. Redenheffer bending awkwardly over the microscope, his tie falling forward into a specimen jar—but when magnification is increased so that the biochemistry of the cell becomes palpable, a brand-new, bright new world appears, the slime now nothing more than the incidental and carelessly contrived canopy that the cell has thrown over itself to disguise from prying eyes the elaborate biochemical machinery that carries on beneath the tent, beautiful as a Balanchine ballet is beautiful, and infinitely more mysterious. The biochemistry has a strikingly abstract and literary nature, so that poor Mr. Redenheffer, had he looked more intently into that microscope, might have seen, staring back at him from the stained plate, a message, some form of meaning, some secret sign of intelligence.

The organization of the bacterial cell is, in fact, like some wonderful Sherlock Holmes mystery in which gradually the superbly anorexigenic Holmes discerns that the cell's bright movements proceed from a central source. The masterful molecule that Mr. Redenheffer did not see and could not imagine—he died in 1956, his obituary revealing to any number of astonished high-school students that he had served with the OSS and parachuted into occupied France—is alive and quivering with information, the place where the body plan of an organism is written down. It is DNA, of course, the great molecular specter.

DNA is a double helix, this everyone now knows, the image as familiar as Marilyn Monroe, two separate strands linked to one another by a succession of steps, so that the molecule itself looks like an ordinary ladder seen underwater, the strands themselves curved and waving. Information is stored on each strand by means of four bases—**A** (adenine), **T** (thymine), **G** (guamine), and **C** (cytosine); these are by nature chemicals (*bases,* of course, and so proton donors), but they function as *symbols,* the instruments by which a genetic message is conveyed. The bases are grouped into wordlike triplets: **AGC, CTA, TGC, CGA, AGC.** And of these, there are sixty-four.

Now step back for a moment—*please.* The fundamental act of biological creation, the most meaningful of moist mysteries along the great manifold of moist mysteries, is the construction of an organism from a single cell. Look at it backward so that things appear in reverse (I am giving you my own perspective): Viagra discarded, hair returned, skin tightened, that unfortunate marriage zipping backward, teeth uncapped, memories of a radiant young woman running through a field of lilacs, a bicycle with fat tires, skinned knees, Kool-Aid, and New Hampshire afternoons; but where memory fades in a glimpse of the noonday sun seen from a crib in winter, the biological drama only *begins,* for the rosy, fat, and cooing creature loitering at the beginning of the journey, whose existence I am now inferring, the one improbably responding to *kitchy, kitchy coo*, has come into the world as the result of a spectacular nine-month adventure, one beginning with a spot no larger than a pinhead and passing by means of repeated but controlled cellular divisions into an organism of ramified and intricately coordinated structures, these held together in systems, the systems in turn animated and controlled by a rich biochemical apparatus, the process of biological creation like no other seen *anywhere* in the universe, strange but disarmingly familiar, for when the details are stripped away, the revealed miracle seems cognate to miracles of another kind, those in which something is read and as a result something is understood, something is commanded and as a result something is done, something is asked and as a result something is accomplished.

The schedule by which this spectacular nine-month construction is orchestrated lies resident in DNA. Such is the common dogma (which is scant reason, I should add, to think it true). *Schedule* is the appropriate word, for while the outcome of the drama is a surprise, the offspring proving to resemble his maternal uncle and his great aunt (red hair, large prominent ears), the process itself proceeds inexorably from one state to the next, and processes of this sort, which are combinatorial (cells divide), finite (they come to an end in the noble and lovely creature answering to my name), and discrete (cells are cells), would seem to be essentially algorithmic in nature, the algorithm now making and marking its advent within the very bowels of life itself.

Now, if DNA functions as life's growing algorithm, another group of molecules—the *proteins*—are its staff; it is there that the genetic message is *expressed*. No surprise, this. The resolution of every organism to its biochemical constituents reveals precisely the same platform: living creatures are constructed of the proteins, which stand to the ant, the field mouse, and the mole as bricks to various buildings.

The proteins are complex molecules, composed in turn of various amino acids. There are twenty such acids in all, and most proteins are chains roughly 250 residues in length. (A residue is an amino acid from which water has been squeezed, leaving behind a dry husk.) This allows life the luxury of choosing from more than 20^{250} possible proteins, the multiplication of possibilities another example of combinatorial inflation, as when routes grow explosively between cities, or sentences accumulate in a natural language.

Some literary modality is yet missing from the equation. A library is in place, one that stores information, and far away, where the organism itself carries on, one sees the purposes to which the information is put, an inaccessible algorithm ostensibly orchestrating the entire affair. But by itself, one biochemical universe, that of the nucleic acids, has nothing to do with the other biochemical universe, that of the proteins. And yet the message written in one is expressed in the other. What is wanted, of course, is a code. A *code,* meaning a conveyance from one world to another. And what is needed, life supplies in the form of the *genetic* code, which serves to associate to each nucleic triplet a corresponding amino acid (see figure 14.2).

It is here that the secrets of creation are partially revealed, for the code shows how triplets on a molecule of DNA are paired, and paired directly, to certain amino acids, the genetic code serving to interpret a command in one language and thus establish that it is expressed in another. The code works by the simplest of stratagems. DNA is read, triplet by triplet, so that the order of the triplets induces a corresponding order in the amino acids, until the genetic apparatus is emptied of the whole of its message and a universe of proteins formed. Like the intercity schedule, which is a pale imitation of creation's central fire, the genetic code brings about an associative network of possibilities. It is very nearly universal, the subtle controlling schedule in the life of every living creation; and it is very nearly coextensive with life itself, the message and its meaning arising at the same time four billion or so years ago.

There are in living creatures two universes embodied in two sets of strings. Meaning is inscribed in molecules and so there is something that reads and something that is read; but they are, those strings, richer by far than the richest of novels, for while Tolstoy's *Anna Karenina* can only suggest the woman,

first letter	second letter: U		C		A		G		third letter
U	UUU	phe	UCU	ser	UAU	tyr	UGU	cys	U
	UUC		UCC		UAC		UGC		C
	UUA	leu	UCA		UAA	stop	UGA	stop	A
	UUG		UCG		UAG	stop	UGG	trp	G
C	CUU	leu	CCU	pro	CAU	his	CGU	arg	U
	CUC		CCC		CAC		CGC		C
	CUA		CCA		CAA	gln	CGA		A
	CUG		CCG		CAG		CGG		G
A	AUU	ile	ACU	thr	AAU	asn	AGU	ser	U
	AUC		ACC		AAC		AGC		C
	AUA		ACA		AAA	lys	AGA	arg	A
	AUG	met	ACG		AAG		AGG		G
G	GUU	val	GCU	ala	GAU	asp	GGU	gly	U
	GUC		GCC		GAC		GGC		C
	GUA		GCA		GAA	glu	GGA		A
	GUG		GCG		GAG		GGG		G

14.2. *The genetic code. The amino acids represented by the sixty-four possible base triplets, or codons, of mRNA. The RNA codons can be directly linked to DNA codons through base-pairing rules, but it is the RNA codons that are directly translated during protein synthesis. Here, U is uracil, which replaces DNA's T (thymine) in RNA. Helena Curtis,* Invitation to Biology, *4th ed. (New York: Worth Publishers, 1985): 194.*

her black hair swept into a chignon, the same message, carrying the same meaning, when read by the right biochemical agencies, can bring the woman to vibrant and complaining life, reading now restored to its rightful place as a supreme act of creation.

Presiding mournfully over his evil-smelling beakers and flasks, his strangely orange toupee askew and his tie stained bright yellow, was this something that Mr. Redenheffer might ever have foreseen?

VOICE OF FOREVER

DNA is a molecule involved in creation, those triplets storing up information and then discharging what they have stored; but DNA is also a molecule meant for immortality, the molecule's design shaped to allow it to penetrate the future. The mechanism is simple, lucid, compelling, extraordinary. In transcription, the molecule faces outward to control the proteins. In replication, it is the *internal* structure of DNA that conveys secrets, not from one molecule to another, but from the past into the future.

DNA is double-stranded and each strand in a single molecule is a system for storing information. A scheme of chemical attraction is at work between strands; indeed, some scheme must be at work if the strands are to remain fastened to one another. The sides of an ordinary wooden ladder are held together by the intervening struts. The system as a whole is entirely mechanical. In the case of DNA, the scheme is entirely chemical. Bases on each strand are matched, so that A is attracted to T, and C to G. When I say attracted, I mean that there is an attractive force between A and T, and between C and G, but *not* between A and G. It is this force that binds base pairs together. The binding of base pairs forms the bridge between strands and gives to the doubled coil its ladderlike appearance (see figures 14.3 and 14.4).

At some point in the life of a cell, double-stranded DNA is cleaved, so that instead of a single ladder, two separate strands may be found waving gently, like fronds of seaweed, the bond between base pairs broken. As in the ancient stories in which human beings originally were hermaphroditic, each strand finds itself longingly incomplete, its bases unsatisfied because unbound. In time, bases attract chemical antagonists from the ambient broth in which they are floating, so that if a single strand of DNA contains first A, and then C, chemical

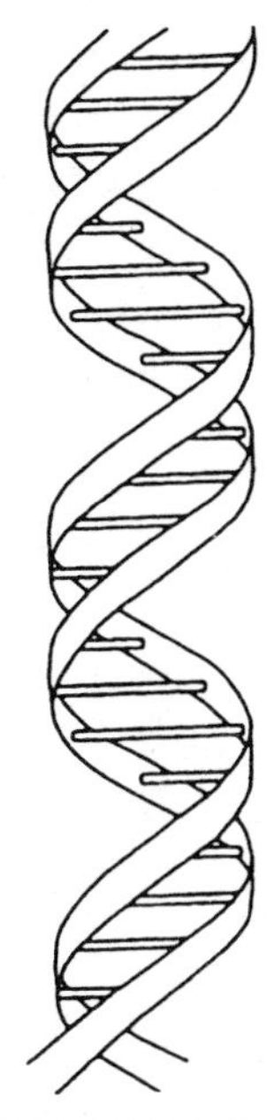

14.3. *DNA helix.*

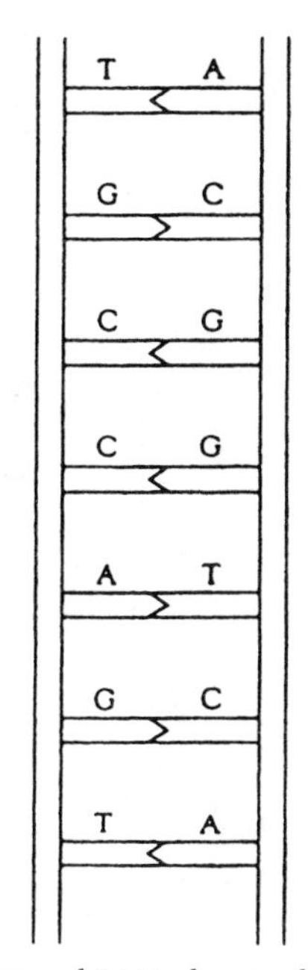

14.4. *Flattened DNA showing base pairs.*

activity, and chemical activity alone, prompts a vagrant T to migrate to A, and ditto for G, which moves to C, so that ultimately, the single strand acquires its full set of complementary base pairs. Where there was only one strand of DNA, there are now two. Naked but alive, the molecule carries on the work of humping and slithering its way into the future.

YES, BUT

Let me ask you something. There really isn't any reason, is there, that biochemical symbols, no less than words in a natural language, could like a searchlight have their focus changed, a particular sequence of bases—C-A-G-A-C-T, say—standing for a city or a town instead of two amino acids? I am drawing the curtain of a connection between a sunbathing Leonard Adelman in sunny Southern California and Arthur Allan Waterman, stamping his feet and freezing his behind before the intercity schedule in hyperborean Witten.

Symbols are symbols, right? And if symbols *are* symbols, then each of the seven cities on Waterman's route can be given a unique biochemical name:

Witten = **ACGATC**

Wittless = **CGCTCA**

Grainball City = **TTGACC**

Sac City = **AGCATA**

Amblot = **AGACCA**

Waterloo = **TACAGG**

Wapping Falls = **TTGAAT**

This scheme serves to associate seven placid midwestern cities with a set of otherwise occupied biochemical bases, an improbable connection forged between very different forms of life. But the scheme does more. It suggests as well a secondary system by which *paths* between cities may *also* be biochemically represented, the path between Witten and Wittless formed by the trick of piecing together the *last* three letters in the chemical symbol for Witten and the *first* three letters in the chemical symbol for Wittless: thus **ATCCGC**, a secret but nonetheless sensible name designating the path taken by a yellow interurban bus. The coding is nothing more than another convenient trick, one that may also be accomplished by means of the city names themselves, the same path between Witten and Wittless designated by **TenWit** in ordinary English letters.

A modest background shuffle suffices to prompt a reconfiguration of the intercity schedule itself (see figure 14.5).

Departs	Arrives						
	Witten ACGATC	Wittless CGCTCA	Grainball City TTGACC	Sad Sac AGCATA	Amblot AGACCA	Waterloo TACAGG	Wapping Falls TTGAAT
Witten ACGATC	X	TenWit ATCCGC	TenGrain ATCTTG	X	TenAmb ATCAGA	X	TenWap ATCTTU
Wittless CGCTCA	LessWit TCAACG	X	LessGrain TCATTG	LessSad TCAAGC	X	X	LessWap TCATTG
Grainball City TTGACC	CityWit ACCACG	CityWit ACCCGC	X	X	X	CityWat ACCAGG	CityWap ACCTTG
Sad Sac AGCATA	SacWit ATAACG	X	X	X	SacAmb ATAAGA	SacWat ATATAC	X
Amblot AGACCA	BlotWit CCAACG	BlotWit CCACGC	X	X	X	X	X
Waterloo TACAGG	X	X	X	LooSad AGGAGC	LooAmb AGGAGA	X	X
Wapping Falls TTGAAT	X	X	FallsGrain AATTTG	FallsSad AATAGC	X	X	X

14.5. *Intercity transportation schedule with English and chemical names to indicate routes.*

And with the reconfiguration complete, the first step in a curious process of intellectual subversion has been undertaken.

EXPLANATIONS AT LUNCH

■ ■ ■ ■ I have been explaining all this to my editor and my agent. "It's obvious, right?" I say hopefully.

"Well, yes," my editor says dubiously. She is a handsome woman with high coloring.

We are lunching at a San Francisco restaurant called Jessica's Waddle, one of those odd places in which the mildly marvelous food is displaced in the diner's attention by the service, which lurches from the familiar to the bumbling and back again to the familiar.

"You mean you didn't get the stewed onions?" our waitress asks, her tattoos dark blue in the noonday sun, as she peers intently at the leftover luncheon plates. "Gee."

My agent lifts her lovely face and cups her chin between her thumb and forefinger. "It's very interesting and all, but what does it *mean*?"

"It *means*," my editor says decisively, "that there isn't any difference between ordinary names and the whatever they are, those chemicals. They're both *symbols*. That is what you're saying, isn't it, David? Why am I even asking? *Of course* it's what you're saying."

It *was* what I was saying—I seem to have been elbowed from my own book by my heroines—but what is extraordinary (I am trying to volley these remarks backward) is not so much the symbol system but the discovery of an algorithmic scheme on alien shores.

I say, "I think it was Leibniz who remarked that good symbolism is an invaluable aid to thought."

"That is *exactly,*" my agent says dramatically, "but *exactly,* the sort of thing *not* to put in the book. I mean who is even going to *know* who this Leibniz is?"

My editor tilts her knife toward her plate. "I think what Susan means," she says, "is that your readers are really going to want to know if the experiment *worked.*"

"Oh, yes, it worked beautifully. I thought I made that clear."

"Not to me," my editor says. "*How* did it work? Your readers are going to want to know whether this Edelweiss actually solved the problem. I mean you have these biochemical names and all, but so what?"

I say, "That's what I was getting at. Setting up the symbols really was all that it took to solve the problem."

"But you have to *explain* that to your readers, David."

"You have to explain it to *us.*"

It would seem I have to explain it to *everyone.*

I have left poor Waterman standing before the intercity schedule; what he can see and what we should notice is that the schedule itself links pairs of cities and *only* pairs of cities. What he needs is a link that goes beyond the pairs of cities to encompass seven cities in all; this crucial information the schedule does not directly provide. But what Waterman—and anyone else—can do is check any proposed seven-city link to see whether it meets his particular scheduling needs. A seven-city link has a simple symbolic form: it consists of the names of seven cities. Witten—Wittless—Grainball City—Sad Sac—Waterloo—Amblot—Wapping Falls is an obvious example. But so is Witten—Grainball City—Sad Sac—Waterloo—Wittless—Amblot—Wapping Falls. And so are any number of variants of these links; any link going from Witten to Wapping Falls will do, just so long as it connects the remaining cities in a way that makes it possible for Waterman to visit each city just once.

The schedule tells Waterman whether there is a connection between pairs of cities. There *is* a bus going between Witten and Wittless; and a plane going between Witten and Grainball City. On the other hand, no form of public transportation serves to take Waterman from Witten to Sad Sac. Such are the facts of life. In the original schedule (see figure 14.1) the existence of a path between Witten and Wittless is marked by a glyph, one standing for a bus. The reconfigured schedule uses English or biochemical letters to indicate the paths between particular pairs of cities (see figure 14.5). But if a connection between pairs of cities can be indicated by a fusion of their ordinary English names, it may be indicated as well by a fusion of their standard biochemical names.

"How so?" my editor asks.

"Well, we have these names for cities? And we have a way to represent paths between cities. I mean that business with the last syllable and the first syllable?"

Both women nod their heads.

"Give us an example," my agent asks dubiously.

"TenWitt. I explained that already."

"I just need you to go over it again."

"We can also express TenWitt in biochemical terms, right?"

"*I'm* following," my agent says, "but I worry about your *readers*."

"No, they'll catch on, trust me. The biochemical equivalent of TenWitt is just ATCCGC—the last three letters of ACGATC and the first three letters of CGCTCA. Don't forget, ACGATC stands for Witten and CGCTCA stands for Wittless. I mean, look at the schedule."

"Well, yes," my agent says, wrinkling her nose.

"Now going from Wittless to Grainball City, it's the same old, same old. The schedule tells you there's a bus between cities, right?"

A bright bursting "right"—this from both my agent and my editor.

"So instead of the symbol for a bus, we could also write LessGrain, right?"

Another doubled "right."

"And instead of LessGrain, we could also write TCATTG, right?"

Silence now follows.

"No, no, it's easy. CGCTCA is just the biochemical name for Wittless and TTGACC is the biochemical name for Grainball City, so the path between cities is just the last three symbols of CGCTCA and the first three symbols of TTGACC."

My editor holds her fork poised in midair and says, "But that's really very obvious, David."

"Very obvious," my agent adds.

■ ■ ■ ■

VAT OF GOO

At the heart of life, there lies a mystery in which biochemicals transform themselves into symbols and symbols acquire an identity as biochemicals. In the scheme that I have just outlined, CGCTCA is purely a symbol—it *stands* for something, as all symbols do. As it happens, CGCTCA stands for a particular city. But CGCTCA is also a physical object, one taking its place among the world's other physical objects, and so bound by the forces and regularities that shape matter in all of its modes. It is this double aspect between symbols as symbols and symbols as things that like a red line running through white sand is the very mark of the algorithm, the current of its life; and it is this double aspect that serves as well to offer my own poor Waterman a scheme for the resolution of his scheduling problems.

Symbols must now be allowed to recede from the event horizon. What is in place instead is a large vat of goo, the goo consisting of various biochemicals. Included in the goo are all of the biochemical paths between cities. *All* of them, note. These biochemical paths may easily be prepared in the laboratory using very ordinary biochemical techniques.

What else goes into the vat? Other nucleotides, of course, these having their usual properties. And among those properties, there is one that is crucial. Base pairs bind to one another, so that just as in life itself, the vat of goo contains base pairs that under certain circumstances attract one another tightly and hold fast.

■ ■ ■ ■ "Um, yes," my editor says, "but what does that have to do with getting through those seven cities?

"Everything," I say decisively.

"This you have to explain, David," says my agent, who has been eyeing the dessert tray with what seems to me to be frank longing.

"Look, it's really very simple, but it's also kind of hard to see. Suppose you have a path between Witten and Wittless, right?"

Doubled nods now.

"And suppose you also have a path between Wittless and Grainball City."

"So I'm supposing," says my agent, who has turned her face decisively away from the pastries.

"What you need next—what Waterman needs next—is a splint between these two paths."

"A splint?"

"A way of sticking paths together."

"Why do we need that?"

"Because if this vat of goo can put together a splint connecting seven cities, and each of those cities happens to be distinct, then when the splint is *read* it will just be the solution to Waterman's problem."

(If only I had the literary powers to evoke the miracle I am mentioning so casually, the very one suggested by the act of *reading* a string of chemicals.)

I continue nonetheless in plain prose: "That's what I mean by a splint. It's *very* simple. The path from Witten to Wittless is TenWitt or ATCCGC. And the path from Wittless to Grainball City is just LessGrain, or TCATTG.

"You said that already, David," my agent says.

"I know, I just want to make sure it's clear."

"Certainly it's clear," my editor adds.

"All right, if it's clear, then the next step should be clear, too. What we need to do, we need to form a splint from Ten-Witt to LessGrain."

"Why do we need to do that?"—this from both women, at once.

"Because if we can go from TenWitt to LessGrain, then we can tell Waterman that there is a path from Witten to Wittless and then from Wittless to Grainball City."

"So you're saying," my agent says in a burst, "that we need to *splint* Witt to Less. I mean, a splint between Witt and Less is just another way of saying there's a path between Wittless and Grainball City."

"Right."

"But this Waterman needs a path between seven cities," my editor says. "I mean even if you could splint Witten to Grain-ball City, that's only two cities, right?"

"That's right. But it's a start."

My editor and agent sit perched at the edge of their chairs, charming women waiting for the story to continue.

■ ■ ■ ■

And it is within the vat of goo that the story continues, for what the vat of goo manages to do—what it had already done, in fact—is to build paths between paths by nothing more complicated than the mechanisms of attraction and repulsion. Base pairs attract one another, that is their nature. And within the vat, the various possible paths between paths form themselves by means of just this attraction. The path between Witten and Wittless is ATCCGC; the path from Wittless to Grainball City is TCATTG. To splint these paths, CGC must be splinted to TCA. That splinting can be managed by a biochemical six nucleotides in length, whose first three nucleotides bind to CGC, and whose second three nucleotides bind to TCA. In fact, GCGAGT is precisely such a sequence. You must thus imagine ATCCGC adrift in the vat of goo, and so, too, TCATTG. Encountering quite by chance the sequence GCGAGT, they come together and clump, this by means of the fact that GCGAGT holds the two sequences together, like a clasp:

Splint
GCGAGT
| | | | | |
ATCCGCTCATTG
Two Paths

As the various chemicals slosh around, binding occurs by chance, and so, too, splinting; but the important and mysterious fact is that it does occur. Once the process commences, it continues. Paths are formed between paths, and then paths between the resulting paths, until in the end, the vat of goo has produced countlessly many paths between cities, including countlessly many paths between seven cities.

■ ■ ■ ■ "There's nothing hard about this," I say. "It isn't rocket science."

"That's exactly, *but exactly,* the tone you've got to avoid," my agent says.

She sits there like that, somewhat out of breath, a delicate woman with dark fluffy hair and brooding eyes.

"Absolutely," my editor adds at once.

"You see, what Adelman realized, and this is really the clever part, he realized that all he needed was to put together the cities *and* the paths between them *and* the bases that bind to the paths."

"How is everything there, folks?"—this from some twenty-something in chinos and a white shirt who appears to be posing as a headwaiter. We all say "fine" at once. He nods and drifts off.

"Then chemistry took over, these chemicals sort of bump into one another, building up longer and longer paths."

"And you're saying this *worked*?"

"Oh, yes. I mean it took a week, but in the end, Adelman was able to find the right path, the one that went from Witten to Wapping Falls and touched every one of the cities on his route just once."

Our waitress reappears to ask about dessert. "How's the bread pudding?" my editor asks. "I love bread pudding."

"It's kind of gross," our waitress says.

We sit for a few more minutes, waiting for the check, the long shafts of California sunshine spilling onto the floor from the skylight.

"How did this Adelman find the right path?"

"It's easy. You remember what I said about these kinds of problems? It's easy to check a solution, but it's hard to find it? The goo produced a whole batch of possible solutions. All Adelman had to do was kind of filter the goo to see if there was one solution that met Waterman's criterion. He just used standard biochemistry to sort of search through the paths and see if one of them went from Witten to Wapping Falls passing every other city just once. It took him a week, but he found it."

"I really liked your Arthur Allen Waterman," my editor finally says. "This Edelweiss you keep talking about, he needs more work. I don't think he's very realized."

"I don't see him at all," my agent agreed.

What could I say? I saw him as a slight man of middle height; high, receding sandy blonde hair, swept backward; an expanse of tanned forehead—he is, after all, from the University of Southern California—the face finely made and sharply featured; the restless bright blue eyes alone suggesting by their depthlessness a dark dissatisfaction.

"Are you just making this up, David, or have you actually met the man?"

"Well, more or less making it up."

"More or less?"

"Sort of more."

"How much more?"

"Entirely."

■ ■ ■ ■

ALL THAT GOO

Somewhere, in some laboratory, there must really have been a vat of biochemical goo, the long biochemical molecules bumping into one another *kachunga, kachunga,* until finally, quite by means of chance and the laws of chemistry, the right path formed itself, the one Adelman could discern by means of his ingenious code that represented a seven-city route from Witten to Wapping Falls.

It was the goo, I suppose, that seized my imagination—and everyone else's imagination as well. Adelman's experiment was widely reported, and like so many other biological experiments, everyone appreciated its success without quite understanding its significance.

Nothing in the goo changed the fundamentals, of course; the goo managed to implement the same algorithm that a computer might have used, but the algorithm remained resolutely nondeterministic, the goo casting about randomly, just as the algorithm required. No, and it wasn't surprising that the goo managed to solve the problem. There are, after all, 10^{23} molecules in a single tablespoon of water, and we are talking here of a *vat* of goo. It was immensely ingenious, no doubt about that, for Adelman to recognize the rich if latent ability of certain biochemicals to store information, but what was there in the experiment that seemed odd, disorienting, and disturbing?

The dissolution of established categories, perhaps, a kind of alteration in expectation; *no, no,* more than that, it was the appearance of *intelligence* on alien shores that seemed to set the nerve of intuition quivering. The dark mysterious goo has been a fixture in our collective imagination for many years, the stuff a staple in the kind of film in which a scientist peers intently into a beaker and then exclaims to his assistant, "*My God, it's growing!*" but in Southern California, Adelman actually got the goo to *do* something and so revealed that, in that phantom molecular world, orders may be given *and* obeyed, the flicker of alertness in the goo suggesting that whenever predictable, organized reactions are possible, so, too, computations, so, too, intelligence, so, too, mind.

But the metaphor is more suggestive than anything it might mean, and in the end the mind that is evident in the goo is one reflectively bouncing back to Adelman himself. And so an old familiar mystery returns to haunt the scene. How does intelligence gain ascendancy over matter?

How *does* it?

CHAPTER 15

The Cross of Words

For the moment, we are all waiting for the gates of time to open. The heroic era of scientific exploration appears at an end, the large aching questions settled, the physical nature of reality understood. Surveying the universe from the place where space and time are curved to the arena of the elementary particles, physicists see nothing but matter in one of its modes. Everything else is either an aspect of matter or an illusion. It is hardly an inspiring view. And few have been inspired by it. "The more the universe seems comprehensible," Steven Weinberg has written sourly, "the more it seems pointless." It is perhaps for this reason that ordinary men and women regard scientific thought with frank loathing. Yet even as the system is said to be finished, with only the details to be put in place, a delicate system of subversion is at work, the very technology made possible by the sciences themselves undermining the foundations of the edifice, compromising its principles, altering its shape and the way it feels, conveying the immemorial message that the land is more fragrant than it seemed at sea.

THE UNDEAD

Entombed in one century, certain questions sometimes arise at the threshold of another, their lunatic vitality strangely intact, rather like one of the Haitian undead, hair floating and silvered eyes flashing. Complexity, the Reverend William Paley observed in the eighteenth century, is a property of things, one notable as their shape or mass; but complexity, he went on to observe, is *also* a property that requires an explanation, some scheme of explication. The simple structures formed by the action of the waves along a beach—the shapely dunes, sea caves, the sparkling pattern of the perishable foam itself—may be explained by a lighthearted invocation of air, water, and wind; but the things that interest us and compel our fascination are different. The laws of matter and the laws of chance, *these,* Paley seemed to suggest, control the behavior of ordinary material objects; but nothing in nature suggests the appearance of a complicated artifact. Unlike things that are simple, complex objects are logically isolated, and so they are logically unexpected.

In writing about complexity, Paley offered the examples that he had on hand—a pocket watch, chiefly; but that watch, its golden bezel still glowing after all these years, Paley pulled across his ample paunch as an act of calculated misdirection. The target of his cunning argument lay elsewhere, with the world of biological artifacts: the orchid's secret chamber, the biochemical cascade that stops the blood from flooding from the body at a cut. These, too, are complex, infinitely more so than a watch, and with these extraordinary objects now open for dissection by the biological sciences, precisely the same inferential pattern that sweeps back from a complex human artifact to the circumstance of its design sweeps back from complex biological artifacts to the circumstance of their design.

What, then, is the *origin* of their complexity? This is Paley's question.

It is ours as well.

GARDEN OF THE BRANCHING FORKS

In developing his argument, Paley drew—he intended to draw—a connection between complexity and design and so between complexity and intelligence. Whether inscribed on paper or recorded in computer code, a design is, after all, the physical overflow of intelligence itself, its trace in matter. A large, general biological property, intelligence is exhibited in varying degrees by everything that lives. It is intelligence that immerses living creatures in time, allowing the cat and the cockroach alike to peep into the future and remember the past. The lowly paramecium is intelligent, learning gradually to respond to electrical shocks, this quite without a brain, let alone a nervous system. But like so many other psychological properties, intelligence remains elusive without some public set of circumstances to which one might point with the intention of saying, "*There,* that is what intelligence *is,* or what intelligence is *like.*"

The stony soil between mental and mathematical concepts is not usually thought efflorescent, but in the idea of an *algorithm* modern mathematics does offer an obliging witness to the very idea of intelligence. Like almost everything in mathematics, algorithms arise from an old and wrinkled class of human artifacts, things so familiar in collective memory as to pass unnoticed. By now, the ideas elaborated by Gödel, Church, Turing, and Post have passed entirely into the body of mathematics, where themes and dreams and definitions are all immured, but the essential idea of an algorithm blazes forth from any digital computer, the unfolding of genius having passed inexorably from Gödel's incompleteness theorem to Space Invaders VII rattling on an arcade Atari, a progression suggesting something both melancholy and exuberant about our culture.

The computer is a machine and so belongs to the class of things in nature that do something; but the computer is also a device dividing itself into aspects, symbols set into software to the left, the hardware needed to read, store, and manipulate the software, to the right. This division of labor is unique among human artifacts: it suggests the mind immersed within the brain, the soul within the body, the presence anywhere of spirit in matter. An algorithm is thus an ambidextrous artifact, residing at the heart of both artificial *and* human intelligence. Computer science and the computational theory of mind appeal to precisely the same garden of branching forks to explain what computers do or what men can do or what in the tide of time they have done.

An algorithm is a scheme for the manipulation of symbols, but to say this is only to say what an algorithm does. Symbols do more than suffer themselves to be hustled around; they are there to offer their reflections of the world. They are instruments that convey information.

The most general of fungible commodities, information has become something shipped, organized, displayed, routed, stored, held, manipulated, disbursed, bought, sold, and exchanged. It is, I think, the first entirely abstract object to have become an item of trade, rather as if one of the Platonic forms were to become the subject of a public offering. The superbly reptilian Richard Dawkins has written of life as a river of information, one proceeding out of Eden, almost as if a digital flood had evacuated itself at its source. Somewhere in the American Midwest, a physicist has argued for a vision of reincarnation in which human beings may look forward to a resumption of their activities after death, this on the basis of simulation within a gigantic computer.

Claude Shannon's promotion of information from an informal concept to the mathematical Big Time served to clarify a difficult concept; it served another purpose as well, and that is to enable *complexity* itself to be brought into the larger community of properties that are fundamental because they are measurable. The essential idea is due ecumenically to the great Russian mathematician, Andrei Kolmogorov, and to Gregory Chaitin, an American student at City College at the time of his discovery (the spirit of Emil Post no doubt acting ectoplasmically). The cynosure of their concerns lay with strings of symbols. Lines of computer code, and so inevitably algorithms, are the obvious examples, but descending gravely down a wide flight of stairs, black hair swept up and held by a diamond clasp, Anna Karenina and Madame Bovary in the end reduce themselves to strings of symbols, the ravishing women vanishing into the words, and hence the symbols, that describe them.

In such settings, Kolmogorov and Chaitin simultaneously observed a sympathetic current running between randomness and *complexity*. A painting by Jackson Pollock is complex in that nothing short of the painting conveys what the painting itself conveys. Looking at those curiously compelling, variegated, aggressively random streaks and slashes, words fail me—*me! Of all people*. In order to describe the painting, I must exhibit it. This is an extraordinary idea, one that captures something long sensed but never quite specified. The complexity of things is tied to the circumstances of their description. An Andy Warhol painting, by way of comparison, subordinates itself to a trite verbal formula: Just run those soup cans up and down the canvas, Andy. *Attaboy.*

An object is complex if no way short of presenting the object exists in order to convey what the object conveys; it is simple to the extent that it is not complex. This is to remain within a rhetorical circle. Instead of paintings, the mathematician attends to binary sequences—strings of 0s and 1s. A string is simple, Chaitin and Kolmogorov argued, if the string may be generated by a computer program (and so an algorithm) significantly shorter than the string itself, and it is complex otherwise, the randomness of old emerging as a simple, solid measure of complexity. Certain strings of symbols may be expressed, and expressed completely, by strings that are shorter than they are; they have some give. A string of ten *H*s (HHHHHHHHHH) is an example. It may be replaced by the command, given to a computer, say, to print the letter *H* ten times. The command is shorter than the string. Strings that have little or no give are what they are. No scheme for their compression is available. The obvious example are random strings—HTTHHTHHHT, say, which I have generated by flipping a coin ten times. Kolmogorov and Chaitin identified the complexity of a string with the length of the shortest computer command capable of generating that string. This returns the discussion to the idea of information, which functions in this discussion (and everywhere else) as a massive gravitational object, exerting enormous influence on every other object in its conceptual field.

What lends this definitional maneuver its drama is a double reduction. The information resident in things comes to be represented by strings of binary digits, and their controlling description by a computer program. But a computer program may itself be depicted as a string of symbols. The familiar contents of the universe have now been evacuated, so that randomness, complexity, simplicity, and information play simply over a pit of writhing strings, things implacable as snakes.

To a certain extent, the concept of complexity serves to explain the large irrelevance that envelops the mathematical sciences. The laws of nature constitute a handful of symbols splayed carelessly across the pages of a text, a tight sequence of symbols controlling the far-flung fabric of creation. These compressed and gnostic affirmations pertain to the large-scale structures of space and time, what is far away and remote, and to the jiggling fundaments of the quantum world. It is amidst the very large and the very small that complexity has loosened its hold. The geometry of space and time is simple enough to be studied and so, too, the quantum world. And yet *most* strings and thus most things are complex and not simple. They cannot be more simply conveyed. They are what they are. They remain unsusceptible to compression. And this is an easily demonstrated, an indubitable, mathematical fact. It is a fact that explains why the mathematical sciences must forever withdraw themselves from the obvious and the ordinary, the mathematician and the physicist like prospectors turning over tons and tons of topsoil to uncover only a few scant ounces of gold.

On the other hand, it is the world's complexity that is humanly intriguing. The most interesting structures of all may be found *here,* on this soundlessly spinning planet, another compelling fact, and one that explains why so many of us are prepared to treat those various voyages of discovery featured on public television with considerable indifference. After all, what do we see when with summer reruns on the other channels we look elsewhere? Stars blazing glumly in the night sky, the moons of Jupiter, hanging there like testicles, clouds of cosmic dust, an immensity of space, the spare but irritating soundtrack suggesting only the infernal chatter of background radiation.

We live within the confines of our own canvas, like a fly trapped inside a Pollock painting. The future that we contemplate is much shorter than the predictable future of the mathematical sciences. Complexity is everywhere, whether created or contrived, and compression hard to come by—in truth the human world cannot be much compressed at all. The most we can typically do, a few resolute morals or maxims aside, is *watch* the panorama unfold itself, surprised as always by the complex, turbulent, and unsuspected flow of things, the gross but fascinating cascade of life.

The definition of algorithmic complexity gives the appearance of turning the discussion in a circle, the complexity of things explained by an appeal to intelligence, and intelligence explained, or at least exhibited, by an appeal to a community of concepts—*algorithms, symbols, information*—upon which a definition of complexity has been impressed. In fact, I am not so much moving aimlessly in a circle as descending slowly in a spiral, explaining the complexity of things by an appeal to the complexity of strings. *Explanation* is, perhaps, the wrong, the dramatically inflated, word. Nothing has been explained by what has just been concluded. The alliance between complexity and intelligence that Paley saw through a dark glass—that remains in place; but the descending spiral has done only what descending spirals can do, and that is to convey a question downward.

A STATION OF THE CROSS

Molecular biology has revealed that whatever else it may be, a living creature is also a combinatorial system, its organization controlled by a strange text, hidden and obscure, written in a biochemical code. It is an algorithm that lies at the humming heart of life, ferrying information from one set of symbols (the nucleic acids) to another (the proteins). An algorithm? How else to describe the intricacy of transcription, translation, and replication than by an appeal to an algorithm? For that matter, what else to call the quantity stored in the macromolecules other than information? And if the macromolecules store information, they function in some sense as symbols.

We are traveling in all the old, familiar circles. Nonetheless, molecular biology provides the first clear, the first resonant, answer to Paley's question. The complexity of human artifacts, the things that human beings make, finds its explanation in human intelligence. The intelligence responsible for the construction of complex artifacts—watches, computers, military campaigns, federal budgets, this very book—finds *its* explanation in biology. This may seem suspiciously as if one were explaining the content of one conversation by appealing to the content of another, and so, perhaps, it is, but at the very least, molecular biology represents a place lower on the spiral than the place from which we first started, the downward descent offering the impression of progress if only because it offers the impression of movement.

However invigorating it is to see the threefold pattern of algorithm, information, and symbol appear and reappear, especially on the molecular biological level, it is important to remember, if only because it is so often forgotten, that in very large measure we have no idea how the pattern is amplified. The explanation of complexity that biology affords is yet largely ceremonial. A living creature is, after all, a concrete, complex, and autonomous three-dimensional object, something or someone able to carry on its own affairs; it belongs to a world of its own—*our* world, as it happens, the world in which animals hunt and hustle, scratch themselves in the shedding sun, yawn, move about at will. The triple artifacts of algorithm, information, and symbol are abstract and *one-*

dimensional, entirely static; they belong to the very different universe of symbolic forms. At the very heart of molecular biology, a great mystery is vividly in evidence, as those symbolic forms bring an organism into existence, control its morphology and development, and slip a copy of themselves into the future. The transaction hides a process never seen among purely physical objects, although one that is characteristic of the world in which orders are given *and* obeyed, questions asked *and* answered, promises made *and* kept. In that world, where computers hum and human beings attend to one another, intelligence is always relative to intelligence itself, systems of symbols gaining their point from having their point gained. This is not a paradox. It is simply the way things are. Two hundred years ago, the Swiss biologist Charles Bonnet—a contemporary of Paley's—asked for an account of the "mechanics which will preside over the formation of a brain, a heart, a lung, and so many other organs." No account in terms of mechanics is yet available. Information passes from the genome to the organism. Something is given and something read; something ordered and something done. But just who is doing the reading and who is executing the orders, this remains unclear.

THE COOL CLEAN PLACE

The triple concepts of algorithm, information, and symbol lie at the humming heart of life. How they promote themselves into an organism is a part of the general mystery by which intelligence achieves its effects. But just how in the scheme of things did these superb symbolic instruments come into existence? Why should there be very complex informational macromolecules at all? We are looking farther downward now, toward the laws of physics.

Darwin's theory of evolution is widely thought to provide purely a materialistic explanation for the emergence and development of life; but even if this extravagant and silly claim is accepted at face value, no one suggests that theories of evolution are in any sense a fundamental answer to Paley's question. One can very easily imagine a universe rather like the surface of Jupiter, a mass of flaming gases, too hot or too insubstantial for the emergence or the flourishing of life. There is in the universe we inhabit a very cozy relationship between the fundamental structure of things and our own boisterous emergence on the scene, something remarked by every physicist. The theory of evolution is yet another station of the cross, a place to which complexity has been transferred and from which it must be transferred anew.

The fundamental laws of physics describe the cool clean place from which the world's complexity ultimately arises. What else besides those laws remains? They hold the promise of radical simplicity in a double sense. Within the past quarter century, physicists have come to realize that theories of change may always be expressed in terms of the conservation of certain quantities. Where there is conservation, there is symmetry. The behavior of an ordinary triangle in space preserves three rotational symmetries, the vertices of the triangle simply changing positions until the topmost vertex is back where it started. And it preserves three reflectional symmetries as well, as when the triangle is flipped over its altitude. A theory of how the triangle changes its position in space is at the same time—it is one and the same thing—a theory of the quantities conserved by the triangle as it is being rotated or reflected. The appropriate object for the description of the triangle is a structure that exhibits the requisite symmetry. Such are the groups. The fundamental laws of physics—the province of *gauge* theories—achieve their effect by appealing to symmetries. And so the physicist looks upon a domain made simple because made symmetrical. This sense of simplicity is a sense of simplicity in things; whether captured fully by the laws of nature or not, symmetry is an objective property of the real world.

The laws of nature are radically simple in a second sense. They are, those laws, simple in their structure, exhibiting a shapeliness of mathematical form and a compactness of expression that itself cannot be improved in favor of anything shapelier or more compact. They represent the hard lump into which the world of matter has been compressed. This is to return the discussion to symbols and information, the cross of words throwing a queer but illuminating lurid red light onto the laws of physics.

The fundamental laws of physics capture the world's patterns by capturing the play of its symmetries. Where there is pattern and symmetry, there is room for compression, and where room for compression, fundamental laws by which the room is compressed. At the conceptual basement, no further explanation is possible. The fundamental laws are simple in that they are *in*compressible; it is for this reason that they are short—astonishingly so in that they may be programmed in only a few pages of computer code.

It is with the fundamental laws of physics that finally Paley's question comes to an end; it is the place where Paley's question must come to an end if we are not to pass with infinite weariness from one set of complicated facts to another. It is thus crucial that they be simple, those laws, so that in surveying what they say, the physicist is tempted no longer to ask for an account of *their* complexity. Like the mind of God, they must explain themselves. At the same time, they must be complete, explaining everything that is complex. Otherwise what is their use? And, finally, the fundamental laws must be material, offering an account of spirit and substance, form and function, all of the insubstantial aspects of reality, in terms (metaphorically) of atoms and the void. Otherwise they would not be fundamental laws of physics.

THE INFERENTIAL STAIRCASE

Triage is a term of battlefield medicine. That shell having exploded, the tough but caring physician divides the victims into those who will not make it, those who will, and those who might. The fundamental laws of physics were to provide a scheme of things at once materialistic, complete, and simple. By now we know, or at least suspect, that materialism will not make it. And not simply because symbols have been given a say-so in the generation of the universe. Just between us, physics is simply riddled with nonmaterial entities: functions, forces, fields, abstract objects of every stripe and kind, waves of probability, the quantum vacuum, entropy and energies, and just recently, mysteriously vibrating strings and branes.

There remains completeness and simplicity. Completeness is, of course, crucial; for without completeness, there is no compelling answer to Paley's question at all; what profit would there be if the laws of physics explained the complexity of plate tectonics but *not* the formation of the ribosomes? Absent completeness, may not the universe break apart into separate kingdoms ruled by separate gods, just as, rubbing their oiled and braided beards, the ancient priests foretold? It is a vision that is at issue, in part metaphysical in its expression, but in part religious in its impulse, the fundamental laws of physics functioning in the popular imagination as demiurges, potent and full of creative power. And if they *are* potent and full of creative power, they had better get on with the full business of creation, leaving piecework to part-time workers.

What remains to be completed for this, the most dramatic of visions to shine forth irrefrangibly, is the construction of the inferential staircase leading *from* the laws of physics *to* the world which lies about us, corrupt, partial, fragmented, messy, asymmetrical, but our very own—beloved and irreplaceable.

No one expects the laws of physics by themselves to be controlling. "The most extreme hope for science," Steven Weinberg admits (I am quoting him for the second time), "is that we will be able to trace the explanations of all natural phenomena to final laws *and* historical accidents." Why not give historical accidents their proper name?—chance. The world and everything in it, the slightly revised Weinberg might have written, has come into being by means of the laws of physics and by means of chance.

A premonitory chill may now be felt sweeping the room. "We cannot see," Richard Feynman wrote in his remarkable lectures on physics, "whether Schrödinger's equation [the fundamental law of quantum mechanics] contains frogs, musical composers, or morality."

Cannot see? These are ominous words. Without the seeing, there is no secular or sacred vision, and no inferential staircase, only a large, damp, unmortgaged claim.

And a claim, moreover, that others have regarded as dubious. "[T]he formation within geological times of a human body," Kurt Gödel remarked to the logician Hao Wang, "by the laws of physics (or any other laws of similar nature), starting from a random distribution of elementary particles and the field, is as unlikely as the separation by chance of the atmosphere into its components."

This is a somewhat enigmatic statement. Let me explain. When Gödel spoke of the "field" he meant, no doubt, the quantum field; Schrödinger's equation is in charge. And by invoking a "random distribution of elementary particles," Gödel meant to confine the discussion to *typical* or *generic* patterns—chance, again.

Under the double action of the fundamental laws and

chance, Gödel was persuaded, no form of complexity could reasonably be expected to arise. This is not an argument, of course; it functions merely as a claim, although one made with the authority of Gödel's genius; but it is a claim with a queer prophetic power, anticipating, as it does, a very specific contemporary argument.

"The complexity of living bodies," Gödel went on to say, Hao Wang listening without comment, "has to be present either in the material [from which they are derived] or in the laws [governing their formation]." In this, Gödel seemed to be claiming without argument that complexity is subject to a principle of conservation, much like energy or angular momentum. Now every human body is derived from yet another human body; complexity is in reproduction transferred from one similar structure to another. The *immediate* complexity of the human body is thus present in the matter from which it is derived; but on Darwin's theory, human beings as a species are, by a process of random variation and natural selection, themselves derived from less-complicated structures, the process tapering downward until the rich panorama of organic life is driven into an *in*organic and thus a relatively simple point. The origin of complexity thus lies—it *must* lie—with laws of matter and thus with laws of physics. Chance, if it plays any role at all, plays only an ancillary one.

In these casual inferences, Gödel was in some measure retracing a version of the argument from design. The manifest complexity of living creatures suggested to William Paley an inexorably providential designer. Cancel the theology from the argument and a connection emerges between complexity and some form of intelligence. And with another cancellation, a connection between complexity and the laws of physics.

And here is the artful, the hidden and subversive, point. The laws of physics are simple in that they are short; they function, they can *only* function, to abbreviate or compress things that display pattern or that are rich in symmetry. They gain no purchase in compressing strings that are at once long and complex—strings of random numbers, for example, the very record in the universe of chance itself. But the nucleic acids and the proteins are precisely such strings. Complexity and randomness are indistinguishable. We have no idea how they arose; filled with disturbing manic energy, they seem devoid of pattern. They are what they are. Their appearance by means of chance is impossible; their generation from the simple laws of physics ruled out of court simply because the simple laws of physics are, indeed, simple.

Gödel wrote well before the structure of the genetic code was completely understood, but time has been his faithful friend (in this as in so much else). In expatiating grandly on biology, Gödel chose to express his doubts on a cosmic scale, his skepticism valuable as much for its sharp suggestiveness as anything else; but the large question that Gödel asked about the discovery of life in a world of matter has an inner voice within biology itself, interesting evidence that questions of design and complexity are independent of scale. Living systems have achieved a remarkable degree of complexity within a very short time, such structures as the blood-clotting cascade or the immune system or human language suggesting nothing so much as a process of careful coordination and intelligent design.

It would seem that in order to preserve the inferential staircase some compromises in the simplicity of the laws of nature might be in prospect. Chance alone cannot do quite what chance alone was supposed to do. Roger Penrose has argued on thermodynamic grounds that the universe began in a highly unusual state, one in which entropy was *low* and organization high. Things have been running down ever since, a proposition for which each of us has overwhelming evidence. This again is to explain the appearance of complexity on the world's great stage by means of an intellectual shuffle, the argument essentially coming to the claim that no explanation is required if only because the complexity in things was there all along.

It is better to say frankly—it is intellectually more honest—that *either* simplicity *or* the inferential staircase must go. If simplicity is to join materialism in Valhalla, how, then, have the laws of physics provided an ultimate answer for Paley's question?

THE ALGORITHM TAKES COMMAND

Although physicists are quite sure that quantum mechanics provides a complete explanation for the nature of the chemical bond (and so for all of chemistry), they are sure of a great many things that may not be so; and they have provided the requisite calculations only for the hydrogen and helium atoms. Quantum mechanics was, of course, created before the advent of the computer. It is a superbly linear theory; its equations admit of exact solution. But beyond the simplest of systems, a computational wilderness arises—mangrove trees, streaming swamps, some horrid thing shambling through the undergrowth, eager to engage the physicists (and the rest of us) in conversation.

Hey, fellas, wait.

The fundamental laws of physics are in control of the fundamental physical objects, giving instructions to the quarks and bossing around the gluons. The mathematician yet rules the elementary quantum world. But beyond their natural domain, the influence of the fundamental laws is transmitted *only* by interpretation.

As complex systems of particles are studied, equations must be introduced for each particle, the equations, together with their fearful interactions, settled and then solved. There is no hope of doing this analytically; the difficulties tend to accumulate exponentially. Very powerful computers are required. The content of the fundamental laws of physics is *relative*—it *must* be relative—to the computational systems needed to interpret them. Thus the advent of the algorithm; thus, its importance; thus the change that it has brought about.

There is a world of difference between the character of the fundamental laws, on the one hand, and the computations required to breathe life into them, on the other. Law and computation have very different natures, very different properties. The laws are infinite and continuous, and so a part of the great tradition of mathematical description that is tied ultimately to the calculus.

Algorithmic computations are neither infinite nor continuous, but finite and discrete. They provide no deep mathematical function to explain anything in nature; they specify only a series of numbers, the conversion of those numbers into a pattern a matter for the mathematician to decide.

If the law describes the hidden heart of things, a simulation provides only a series of stylized snapshots, akin really to the succession of old-fashioned stills that New York tabloids used to feature, the woman, her skirt askew, falling from the sixth-floor window, then passing the third floor, a look of alarm on her disorganized features, finally landing on the hood of an automobile, woman and hood both crumpled. The lack of continuity in either scheme make any interpolation between shots *or* simulation a calculated conjecture. The snapshots, it is worthwhile to recall, provide no evidence that the woman was falling *down*; we see the pictures, we neglect the contribution *we* make to their interpretation.

Computational schemes figure in any description of the inferential staircase; the triple concepts of algorithm, information, and symbol have made yet another appearance. They are like the sun, which comes up anew each day; but *they* serve to sound a relentlessly human voice. The expectation among physicists, of course, is that these concepts play merely an ancillary role, the inferential staircase under construction essentially by means of the fundamental laws of physics; but the question of who is to be the master and who the mastered in this business is anything but clear.

To the extent that the fundamental laws of physics function as premises to a grand metaphysical argument, one in which the universe is to appear as its conclusion, some specification of the triple concepts functions as additional premises. Without those premises, the laws of physics would sit simply in the shedding sun, mute, inglorious, and unrevealing. A third revision of Steven Weinberg's memorable affirmation is now in prospect: the best that can be expected, the most extreme hope for science, is an explanation of all natural phenomena on the basis of the fundamental laws of physics *and* chance *and* a congeries of computational schemes, algorithms, specialized programming languages, techniques for numerical integration, huge canned programs (such as Mathematica or Maple), computer graphics, interpolation methods, computer-theoretic shortcuts, and the best efforts by mathematicians and physicists to convert the data of simulation into coherent patterns, artfully revealing symmetries and continuous narratives.

A certain radiant expectation may now be observed dimming itself. The greater the contribution of an algorithm, the less compelling the vision of an inferential staircase. Matters stand on a destructive dilemma. Without the triple concepts, the fundamental laws of physics are incomplete; but with the triple concepts in place, they are no longer simple. The inferential staircase was to have led from the laws of physics to that sense of intelligence by which complexity is explained; if intelligence is required to *construct* the very stairs themselves, why bother climbing them? The profoundly simple statements that were to have redeemed the world must do their redeeming by means of the very concepts that they were intended to redeem. An algorithm is, after all, an *intelligent* artifact.

We have always suspected, and now we know, that while there are things that are simple and things that are complex in nature, the rich variety of things is not derived from anything simple, or if derived from something simple, not derived from it completely. Complexity may be *transferred*; it may be shifted from theories to facts and back again to theories. It may be localized in computer derivations, the fundamental laws kept simple, or it may be expressed in a variety of principles and laws at the expense of the idea of completeness; but ultimately, things are *as* they are for no better reason than that they are *what* they are.

THE CARDINAL AT DINNER

Dinner was over. The great table had been cleared of plates, each with a blue-and-green Vatican crest, and its heavy napery. The room glowed with a soft, red light thrown from the fluted wall tapers. A warm blue haze from the pre-Revolutionary Cuban cigars we were all smoking drifted upward, collected itself into a little cloud, and then was whisked away by the room's efficient but unobtrusive ventilation system. The cardinal shifted his ample haunches underneath his red robes, the better, he remarked, to ease the flaming sciatica that traveled from his hips to his ankles. He sipped at the Benedictine that his servant had placed before him; it had been laced with a few drops of valerian to ease the discomfort of his back. I could see the great wave of sleepiness that came over him, but with an act of mental discipline that I imagined years had made habitual, he resisted the impulse to close his eyes.

"My dear Professore Dottore," he said suavely, nodding toward a tall man sitting at the table's far end, "it is now that you must tell us how the world began."

The professore dottore, who was, in fact, a powerful and well-known physicist from the University of Turin, cleared his throat and began to speak in a soft, high tenor. He had clearly prepared his remarks well in advance and he said what he had to say in a kind of florid and old-fashioned German very much at odds with the Viennese dialect that the cardinal (and everyone else) spoke. "As you know, Eminence," he said, "it has been mathematical physics that has embodied the largest, the most generous, intellectual gesture of the past five hundred years."

The cardinal nodded gravely, as if to say that that was certainly so.

"Within the scheme that it elaborates, real numbers are assigned to continuous magnitudes. Time and space are given a quantitative skeleton." The professore dottore paused to look at the men sitting around the table and placed his hands before him on the table and pressed them downward; I noticed somewhat to my surprise that they were trembling just slightly.

He said, "It is a characteristic and artful method of investigation that has been developed, one in which things and processes are described by mathematical equations. It has allowed us to survey the universe from the big bang to the end of time."

"That is very elegantly put," said the cardinal, "but I was under the impression that general relativity and quantum mechanics are yet in conflict. These are the two great visions of your science, *no*?"

The professore dottore nodded vigorously. "You are correct, Eminence," he said. "In general relativity, space and time are fused into a manifold. Matter deforms space and time, and space and time influence the behavior of matter."

"And the other," the cardinal asked, "quantum mechanics?"

"It is in certain respects a dark subject, Eminence."

"*Quantum mechanics* a dark subject? It is a good thing that you never take confession, my friend."

A number of men at the table saluted this sally with deep throaty chuckles.

"A figure of speech, Eminence," said the professore dottore. "Instead of particles, there are waves of probability, which launch themselves throughout the whole of space."

"*Ah*," said the cardinal, who had in his student years in Verona in fact studied quantum mechanics under a physicist who had himself studied with Enrico Fermi. "I remember. Waves behave like particles and particles behave like waves, and solitary photons passing through a slit manage to interfere with themselves."

"Yes," said the professore dottore.

"Extraordinary. And to think there are those who make fun of our poor Church because of the miracle of transubstantiation." The cardinal paused to collect himself and then smiled. "Two entirely different disciplines, *no*? And yet you assume that beneath it all the universe is simple."

"We are searching, Eminence, for the final unified theory."

"And when you have found it?"

"The physical aspects of reality will be subordinated to a single law of nature."

"And this great theory, what will it teach us about the universe, its meaning?"

The professore dottore inserted his index finger beneath his shirt collar to loosen it slightly.

"The more the universe seems comprehensible, Eminence," said the professore dottore, "the more it seems to have no meaning."

"And you are persuaded that it does seem comprehensible?" the cardinal asked. He no longer looked the least bit sleepy.

The professore dottore smiled indulgently. "Evidently," he said, "there are the laws of physics."

The cardinal placed his heavy peasant forearms on the table in front of him. "If the universe is comprehensible, surely that is evidence that it is not entirely lacking in meaning, *no*? A meaningless universe would not be comprehensible."

The Professore Dottore for a moment looked down at his hands. When he looked up, he said, "Forgive me, Eminence, but why *should* it have a meaning?"

"My dear Dottore," said the cardinal, "our question is not *why* the universe has a meaning but *whether* it has one. Surely we can agree to leave the why of things in other hands." And here the cardinal pointed slyly toward the ornate ceiling of the dining room.

The professore dottore shrugged his thin shoulders; he seemed perhaps a touch vexed, but his irritation had enforced his self-confidence and when he spoke again he spoke in a calm, clear voice and his hands had ceased to tremble.

"As far as we can tell, Eminence, the universe is just a physical system. Nothing more."

"A physical system," said the cardinal, "one entirely explained by the laws of matter."

"Entirely, Eminence."

The cardinal nodded his heavy head again. "And this system," he said, "how did it come to be?"

The professore dottore now looked directly out toward the end of the table, where the cardinal was sitting. "I am sure Your Eminence knows the answer," he said tentatively.

"Yes, of course, but the others," said the cardinal, gesturing with a sweep of his thick hand toward the men sitting at the table, his ring just catching and holding the light.

The professore dottore nodded vigorously, as if to say that he quite understood the need for explanation, given that many men at the table were specialists in other subjects. "The current theory," he said, "is that the universe erupted into existence some fifteen billion years ago."

"The big bang," said the cardinal, with just the slightest suggestion in his voice that he was conveying an absurdity.

"It is what everything suggests, Eminence, the data, our theories, everything."

"Yes," said the cardinal, "but what lies before the big bang? I am asking in ignorance."

"Nothing."

"Nothing?"

"Nothing."

"But my dear Professore Dottore, surely this troubles the intellect, *no*? There is nothing for all eternity, and then—*poof*—there is something."

The professor dottore shook his head vigorously. "No, no, Eminence," he said. "There was no eternity before. This is a mistake."

"*Ah*," said the cardinal, "eternity a mistake."

"I am talking only of science, Eminence," said the professore dottore apologetically. "Space and time were created with the universe. They have no prior existence."

"Then there is no time *before* this big bang of which you speak?"

"It seems difficult to grasp, Eminence, but just as there is no point north of the North Pole, so there is no time prior to the big bang."

"Then there is nothing to explain the creation of the universe." The cardinal motioned with his hand to the walls of the great room, heavy with lovely silken tapestries in shades of warm vermilion. "All of this, this grandeur," he said, "just happened for no reason whatsoever? Is that what you believe?"

"No, Eminence," said the professore dottore, "I believe that ultimately the laws of physics explain the existence of the world."

The cardinal shifted his considerable bulk backward, so that he could withdraw his forearms from the table and rest them on the carved wooden armrests of his chair.

"This you must explain to me," he said. "The laws of physics, they are symbols, *no*? Things made of man?"

"Well, yes," said the professore dottore, "in a sense that is right. Mathematical symbols. But they are not made by man."

"Who then?"

"I mean," said the professore dottore, "that they are not made at all. They are just what they are."

"*Ah*," said the cardinal, releasing a stream of air over his full lower lip. "They are what they are."

The professore dottore shrugged his thin shoulders as if to say he was helpless before the facts.

"But whatever they are," said the cardinal, "they are not physical objects, these laws, *no*?"

"The laws of physics do not exist in space and time at all. They describe the world, they are not in it."

"Forgive me," said the cardinal, "I thought I heard you say that everything in the world could be explained by the behavior of matter, *no*? It would appear that you meant everything in the world except the reason for its existence."

"Eminence, every chain of explanation must come to an end."

"Is it not convenient that *your* explanations come to an end just before they are asked to explain a very great mystery?"

The professore dottore shrugged his thin shoulders once again. Like all of us, he had been warned that the cardinal was a man of prepossessing rhetorical power, and he was, I imagine, still enough of a Catholic not to wish to press his own arguments.

"It would seem to me," said the cardinal, sipping again at his Valerian-laced Benedictine, "that if you were serious about this search of yours"—and here he paused for emphasis so that the entire table hung on his every word—"that you would be looking for laws that in the end explain themselves as well as everything else."

The professore dottore looked up attentively but said nothing. Neither did anyone else. The blue cigar smoke continued to drift upward, the moment of tension gathered itself and then collapsed, and then the cardinal, with a warm throaty chuckle, said something in his own slangy Italian dialect to the men sitting directly around him, who threw up their hands and laughed; but speaking no Italian myself, I had no idea what he said.

■ ■ ■ ■

And so they have gathered by the gates of time, the quick and the dead and those eager to be born. A cool gray fog is blowing. The ancients tell stories of why time began and how space was curved. The spidery hands of the great clock, measuring millennia instead of minutes, are crawling toward midnight. The logicians have gathered themselves together. Aristotle is there and Abelard, and Frege and Cantor, still dressed in hospital white. Peano and Hilbert are talking to one another; Russell is staring at Gödel, and Gödel is staring into space. Church has risen from his seat; Turing bends low to catch his breath, and Post pats him upon the back. Then as the hinge of the great gate swings open, Gottfried Leibniz appears at last, his luxuriant wig fluttering in the evening breeze, his arms outstretched toward paradise, and as music fills the air, he begins to dance the slow stately dance whose rhythms signify what has been, what is, and what may come.

EPILOGUE

The Idea of Order at Key West

Wallace Stevens

She sang beyond the genius of the sea.
The water never formed to mind or voice,
Like a body wholly body, fluttering
Its empty sleeves; and yet its mimic motion
Made constant cry, caused constantly a cry,
That was not ours although we understood,
Inhuman, of the veritable ocean.

The sea was not a mask. No more was she.
The song and water were not medleyed sound
Even if what she sang was what she heard,
Since what she sang was uttered word by word.
It may be that in all her phrases stirred
The grinding water and the gasping wind;
But it was she and not the sea we heard.

For she was the maker of the song she sang.
The ever-hooded, tragic-gestured sea

Was merely a place by which she walked to sing.
Whose spirit is this? we said, because we knew
It was the spirit that we sought and knew
That we should ask this often as she sang.

If it was only the dark voice of the sea
That rose, or even colored by many waves;
If it was only the outer voice of sky
And cloud, of the sunken coral water-walled,
However clear, it would have been deep air,
The heaving speech of air, a summer sound
Repeated in a summer without end
And sound alone. But it was more than that,
More even than her voice, and ours, among
The meaningless plungings of water and the wind,
Theatrical distances, bronze shadows heaped
On high horizons, mountainous atmospheres
Of sky and sea.

It was her voice that made
The sky acutest at its vanishing.
She measured to the hour its solitude.
She was the single artificer of the world
In which she sang. And when she sang, the sea,
Whatever self it had, became the self
That was her song, for she was the maker. Then we,
As we beheld her striding there alone,
Knew that there never was a world for her
Except the one she sang and, singing, made.

Ramon Fernandez, tell me, if you know,
Why, when the singing ended and we turned
Toward the town, tell why the glassy lights,
The lights in the fishing boats at anchor there,
As the night descended, tilting in the air,

Mastered the night and portioned out the sea,
Fixing emblazoned zones and fiery poles,
Arranging, deepening, enchanting night.

Oh! Blessed rage for order, pale Ramon,
The maker's rage to order words of the sea,
Words of the fragrant portals, dimly-starred,
And of ourselves and of our origins,
In ghostlier demarcations, keener sounds.

ACKNOWLEDGMENTS

I should like to thank my editor, Jane Isay, and my agent, Susan Ginsburg, for suggesting the idea for this book. I should like to thank Rachel Myers for urging me delicately, but with no little firmness, to get rid of what I had first written and make it better. Certain portions of the last two chapters of this book appeared in very preliminary form in *Commentary* and *Forbes ASAP.* I am grateful to those journals for having afforded me the opportunity to explore certain themes and ideas.

INDEX